T0391134

Solar Radiation Data

Solar Energy R&D in the European Community

Series F:

Solar Radiation Data

Volume 1

Scientific editing: W. PALZ
Publication arrangements: D. NICOLAY

Solar Energy R&D
in the European Community

Series F Volume 1

Solar Radiation Data

Proceedings of the
EC Contractors' Meeting held in
Brussels, 20 November 1981

edited by

W. PALZ
Commission of the European Communities

D. REIDEL PUBLISHING COMPANY
Dordrecht, Holland / Boston, U.S.A. / London, England

for the Commission of the European Communities

Library of Congress Cataloging in Publication Data

Main entry under title:

Solar radiation data.

(Solar energy R&D in the European community. Series F ; v. 1)
1. Solar radiation--Congresses. 2. Solar energy--Congresses.
I. Palz, Wolfgang. II. Commission of the European Communities.
III. Series.
QC910.2.S65 333.79'23 81-23441
AACR2

ISBN-13: 978-94-009-7786-0 e-ISBN-13: 978-94-009-7784-6
DOI: 10.1007/978-94-009-7784-6

Organization of the Contractors meeting by
Commission of the European Communities
Directorate-General Research, Science and Development, Brussels

Publication arrangements by
Commission of the European Communities
Directorate-General Information Market and Innovation, Luxembourg

EUR 7668

Softcover reprint of the hardcover 1st edition 1982

Published by D. Reidel Publishing Company
P.O. Box 17, 3300 AA Dordrecht, Holland

Sold and distributed in the U.S.A. and Canada
by Kluwer Boston Inc.,
190 Old Derby Street, Hingham, MA 02043, U.S.A.

In all other countries, sold and distributed
by Kluwer Academic Publishers Group,
P.O. Box 322, 3300 AH Dordrecht, Holland

D. Reidel Publishing Company is a member of the Kluwer Group

C O N T E N T S

ACTION 4 - NEW MEASUREMENTS AND TECHNOLOGIES

INTRODUCTION

The present proceedings report on the coordination meeting of contractors for solar radiation data which was held in Brussels on 20 November 1981.

The programme on solar radiation data is one of the eight sub-programmes of the solar energy research and development programme of the European Communities which was decided by the Council of Ministers in 1979. It is implemented through contracts with all the European meteorological offices in the member countries or other agencies in charge of gathering those data. There are also contracts with industries, universities and other research institutions. Most of the cost not covered by the Commission's budget is supplied by the national solar energy programmes.

The knowledge of extensive and precise solar radiation data in their appropriate form is of fundamental importance for the future of solar energy developments and applications in Europe. Until now those data are not generally available and even worse the meteorological organisations had not really been motivated to measure the solar radiation data as extensively as the solar energy community needs them. The programme aims at the production of more and more reliable data and at the processing of these data in such a way that they become easily accessible. For the latter a methodology is developed systematically which is validated.

It was in the frame of the Commission's solar energy programme that in 1977 activities on a European level were started on a sufficiently large scale to gather the necessary radiation data and to put them into such a form that they can be easily used by the scientists, design engineers and users of solar coversion systems. So far the programme has been very successful indeed by making all meteorological services and other interested parties work together and thus improving the situation in this important sector.

The activities on solar radiation data are broken down into four areas :

Action 1 concerns the calibration of national solar radiation instruments because it is of paramount importance tc make sure that all measured data are accurate and reliable. Meetings have been organised at Carpentras, France, at which all interested parties in the European Community participated and in which national pyrheliometers and subsequently pyranometers were calibrated.

Action 2 concerns the production of reference years which include for every year typical meteorological data as are needed to design solar energy conversion systems and to simulate their performance by means of precise computer models. Data which are currently included are for temperature, global radiation and alternatively sunshine duration. Some reference years include also data on wind. The methods which are currently set up define first twelve typical months from many years of records and link them together into a complete year. Each reference year is thus a collection of weather data for one "typical" year with 8.760 sets of simultaneous weather parameters. It is delivered on a data carrier suitable for computer treatment, normally a magnetic tape. The steps undertaken so far have resulted in the selection of the so-called "Belgian Method" which is recommended for solar energy in building load applications. It is simple and cheap to use. In the coming months and years it is intended to develop reference years for all typical climates in the European Community.

Besides these very precise and comprehensive data which are only of use for those designers and users who have at their disposal a computer, it is also necessary to convert the meteorological data and in particular the radiation data into simpler representations which are of general interest.

Action 3 has been designed with the purpose of developing and publishing atlases and data books from existing data. In the frame of this activity the first atlas for global radiation was already published in 1980*. A new version of this atlas is in preparation and will be published in the course of 1983. It will be based on a denser network within the European Community and will also include data from additional European countries and in particular from the Mediterranean region. Data are also being produced for later publication for (1) global radiation on tilted planes and (2) direct radiation.

Besides these atlases data books will be provided on statistical analysis of meteorological data in particular the frequency distribution radiation intensity and the link between irradiance and other meteorological parameters.

Action 4 is concerned with the improvement of the national radiation networks. It includes the improvement of the existing network, the densification of network for global radiation and the set-up of a lot more stations for direct and diffuse radiation. Right now, the network for global radiation is far from being dense enough; it is planned to have a density of at least one station per 100 x 100 km. In the longer range there is also a need for more measurements on micro-climates which are important for the many coastal and mountainous areas in Europe.

---

* Commission of the European Communities "European Solar Radiation Atlas" Volume 1: Global Radiation on Horizontal Surfaces, published by W. Grösschen-Verlag.

Action 4 includes also special measurements and the development of a methodology to exploit data which are currently provided by meteorological satellites.

A detailed paper on the activity of the programme including, in particular, all publications, is provided in the strategy paper which is attached.

STRATEGY PAPER

EC SOLAR ENERGY R+D PROGRAMME

PROJECT F, SOLAR RADIATION DATA

Outline

(Status as of April 1981)

INTRODUCTION

The Project is sub-divided into 9 Actions. As a rule, Actions include the following 3 phases :

- development of a methodology
- validation of the methodology
- production of data for publication

In addition, Project F provides services to other activities of the EC Solar Energy Programme.

Services to EC Solar Activities

- Helioelectric power plant "Eurelios" (circumsolar radiation);
- Collector testing activities
- Modelling of solar heated houses
- Photovoltaic pilot projects
- Units and symbols for solar energy (draft report ready).

Action 1 : Calibration and characteristics of radiometers

Intercomparison of pyranometers (in cooperation with IEA + WMO)

In June 1981 : International comparison of pyranometers at Carpentras (F)
Action leader : Mr. Plazy (F)

Action 2 : Reference years

Sub-division

- Test reference years (TRY)
- Short reference years (SRY)

Action leader : Mr. Lund (Dk)

Task

Development and production of a library of data tapes; various sites inside the European Community.

Methodology

- TRY : ready
- SRY : Dec. 1981

Validation of methodology

- TRY : ready
- SRY : July 1981

Products

- Brochure of Action 2, giving a general presentation : ready Aug.1981
  Responsible : Mr. Lund, inputs from Mr. Dogniaux
- TRY
  Available for Belgium and Denmark as part of the validation of methodology.
  Responsible for preparation of other TRY : Mr. Dogniaux
  Sites for TRY are defined by Mr. Lund and Mr. Dogniaux by Sept.1981
  All TRY ready : by June 1983
- SRY : by mid-1983, SRY for same location as for TRY
  Algorithms for the production of SRY and some key values

Distribution of products

From national institutions (No central library)

Action 3 : Atlas and data books (from existing data)

Action 3.1 : Global irradiance on horizontal planes

Definition of radiation climate zones in the EC

Action leader : Dr. Kasten (D)

Tasks

- Improvement and extension of the first volume of the EC solar radiation atlas published in 1980 (in cooperation with the national met. services);
- Definition of radiation climate zones in the EC :
  Mr. Dogniaux provided an "ad hoc" definition in June 1980; work to be finalised by Dec. 1982.

Products

Atlas and data tables for 10 year averages.

July 1981 : Filling in of gaps in the existing atlas (responsible : Mr. Kasten).
New data for all Mediterranean countries.
June 1982 : Data for other neighbouring countries e.g. Austria and Switzerland and Northern countries (responsible : Mr. Kasten)
Dec. 1982 : Final report ready
Negotiations with WMO to become involved in the publication
June 1983 : New atlas ready for printing

Action 3.2 : Global irradiance on tilted planes

Direct radiation

Action leader : Mr. Dogniaux (B)

Tasks

- Calculation of global irradiance on planes oriented south at the angle of the latitude and vertical planes oriented east, south, west; monthly means of daily values (hourly values at a later stage);
- Calculation of direct radiation on horizontal planes (tilted planes at a later stage); hourly values; calculation of the turbidity factor.

Methodology

- Global irradiance on tilted planes by July 1981 (Mr. Page, Mr. Krochmann)
- Direct radiation, by June 1982

Products

- Global irradiance on tilted planes
  July 1981 data for 56 stations (responsible Mr. Krochmann, Mr. Page)
  Oct. 1981 report on the method
  Dec. 1982 data for extended number of stations
- Direct radiation, June 1983, data produced, atlas ready for printing

Action 3.3 : Statistical analysis of meteorological data

Action leader : Mr. Bedel (F)

Tasks

- Cumulative frequency distribution
- Time sequence of irradiation
- Correlation between irradiance and other meteorological parameters e.g. temperature

Methodology

- Cumulative frequency distribution;
  horizontal planes, available
  for hourly data, inclined surfaces, Dec. 1982
- Time sequence of irradiation, available
- Correlation with meteorological parameters, Dec. 1982

Products

- Cumulative frequency distribution, daily values, horizontal planes, for 49 stations, Nov. 1980 (Mr. Kasten)

- Time sequence of irradiation, daily values, Horizontal planes, for 49 stations, Nov. 1980 (Mr. Kasten)
- Time sequence of irradiation, daily values, inclined planes, for 49 stations, Dec 1982 (Mr. Kasten)
- Correlation between irradiation and temperature, June 1983 (Mr. Kasten)

Action 3.4 : Sensitivity analysis of useful energy output for solar converters with respect to quality and completeness of meteorological data sets

Action leader : Mr. Pilatte (B)

Tasks

Guidance to adapt the meteorological measurements to the needs of solar energy users
- re-correlation for reduction method of meteorological data
- recommendation for calculation methods to be used with reduced data

Methodology

Development of appropriate models of reduction July 1981
Application to flat plate solar collector Dec. 1981
Application to other types of collectors Dec. 1982

Products

June 1983 - recommended method of reduction
- utilization procedure of reduced data

Action 4 : New measurements and technologies

Action 4.1 : Improvement of measurements in national networks (global, direct, diffuse irradiance and irradiance on inclined surfaces)

Action leader : Mr. Grüter (D)

Tasks

- Improvement of instrumentation (in particular for diffuse radiation)
- Densification of networks for global, direct and diffuse irradiance
- Measurements on inclined surfaces

Methodology

- Study of the spatial correlating of irradiance for the assessment of necessary density of stations, Dec. 1982
- Harmonisation of measuring techniques on diffuse radiation

Products

- Densification of radiometric networks on global, diffuse and direct radiation
- Recommended method for shade ring measurements
- Data for irradiance on inclined surfaces

Action 4.2 : Special measurements

Action leader : Mr. Plazy (F)

Tasks

Development of instrumentation for specific needs in solar energy
Turbidity, solar aureole, spectral and long wave radiation, distribution of the luminance

Products (Reports)

- Infra-red radiation June 1982
- Turbidity Dec. 1982
- Spectral radiation March 1981
- Circumsolar radiation June 1983

Action 4.3 : Satellite images

Action leader : Mr. Poisson (F)

Task

Development of operational methods using satellite images for the determination of cloudiness, sunshine duration and irradiance

Product

Operational method Dec. 1982

SUMMARY OF PUBLICATIONS (dates refer to when ready for printing)

1. Atlas and data books

Volume 1 (new version) June 1983

- global radiation on horizontal surfaces
- report on climatic zones

Volume 2 Octobre 1981

- global radiation on tilted surfaces (for 49 stations in the EC)
- cumulative frequency distribution for horizontal surfaces (daily values, for 49 stations)
- time sequence of irradation (for 49 stations).

Volume 3 June 1983

- direct radiation

Note : the list of units and symbols is attached to each volume

2. Special reports

- Brochure on reference years, Aug. 1981 (Mr. Lund)
- Report on cumulative frequencies, hourly data, horizontal and inclined surfaces, Dec. 1982 (Mr. Bedel)
- Report on correlation between irradiance and other meteorological parameters, June 1983 (Mr. Bedel)
- Manual to the user of meteorological data, June 1983(Mr.Pilatte)
- Report on IR radiation, turbidity, spectral radiation, circumsolar radiation, June 1983 (Mr. Plazy)

3. Other products

- Tape of test reference years, June 1983
- Operational method to exploit satellite images for the measurement of irradiance, sunshine duration and cloudiness, Dec. 1982

ACTION 1

CALIBRATION AND CHARACTERISTICS OF RADIOMETERS

Action leader : J.-L. PLAZY

Commissariat à l'Energie Solaire, Sophia Antipolis, Valbonne, France

Action 1 - CALIBRATION AND CHARACTERISTICS OF RADIOMETERS

Action leader : M. J.-L. PLAZY
Commissariat à l'Energie Solaire
Sophia Antipolis
F - 06560 VALBONNE

Participants : Direction de la Météorologie — Contrat ESF-005 F
M. P. GREGOIRE

University College of Cardiff
Dr. J. McGREGOR

Tasks :
- Intercomparison of pyranometers
- Characterisation of pyranometers
- Handbook on the use of pyranometers

In order to check the accuracy of the radiation measurements practised in the EC countries, a substandard pyranometer comparison was organized in Carpentras in June 1981. Each National Meteorological Service was requested to send one or two of the pyranometers used by it to calibrate the network pyranometers by a comparison method. These instruments have been compared against a fully characterized reference pyranometer. The results show a difference of about 9 % between the lowest and the highest instruments. In fact, it appeared that some pyranometers were used with the manufacturer's coefficients. the use of the proposed coefficients should reduce the spread of the values to about two per cent. This comparison shows the necessity to use a specified calibration methodology for the transfer instruments.

A better knowledge of the pyranometers' response has been obtained by laboratory tests in order to estimate the possible accuracy which can be achieved in field measurements.

I. INTRODUCTION

In order to ensure comparability of the radiation data for the network measurements and also the calibration methods, a primary national pyrheliometer comparison was held in Carpentras in 1978 which demonstrated a good homogeneity of the standards. In spite of these results it appeared that some differences still existed at the network level in the radiation measurements. One problem is the transfer of the radiometric scale from the pyrheliometers to the pyranometers. An inquiry made through the national meteorological services showed that different methods were employed for the network pyranometer calibration, one for each country.

For this reason some work has been done, first to characterize the pyranometers as well as possible to ascertain the accuracy which it is possible to obtain with such instruments, secondly to develop a calibration methodology to reduce the loss of accuracy during the transfer operations. The pyranometer characterization is being carried out by the Solar Department of the University College in Cardiff and a comparison of substandard transfer pyranometers was held in Carpentras in June 1981.

II. DESCRIPTION OF THE METHODOLOGY AND THE EQUIPMENT

Each country of the EC was asked to send one or two pyranometers, normally used as substandard to calibrate their network pyranometers, to the radiometric centre of Carpentras. With the exception of Italy, each country responded to this request and the comparison was held from June 15 to 25 and early in September for the Danish pyranometer.

During this period there were some cloudless days with high or moderate temperatures, some cloudy days, and one overcast day.

All the pyranometers were connected to an automatic data acquisition system, type INF 128 AERAZUR, with a multi 20 Intertechnique computer. The peripheral equipment comprised a fast printing unit, a mini K7 recorder and a tape drilling machine. Each low level card was connected to 6 pyranometers and a voltage standard in order to monitor and correct a possible drift in the measurements. A temperature input made it possible to correct the values according to the temperature curve for each instrument.

The resolution of the data acquisition system was 10 μV and the accuracy 30 μV.

The radiation value calculated for each pyranometer was determined by using the calibration factor given by the Meteorological Service which owned the instrument, taking into account the temperature correction. In fact for some pyranometers the calibration factor communicated to the Direction de la Météorologie was the calibration factor estabished by the manufacturer.

There were two reference values for the calculation :

reference 1 : values given by the Eppley PSP pyranometer 16542 F 3 using a calibration factor established from numerous series of calibration tests against the sun.

reference 2 : values given by the same pyranometers but using a calibration function and not a calibration factor.

The difference in the daily amount of radiation calculated by the two methods lies between 0.6 and 1.6 %, according to the weather conditions.

The ratios between each pyranometer and the references will be calculated each hour, using ten values taken at 6-minute intervals, each day and for the whole period.

In parallel with this action, two pyranometers previously calibrated in Carpentras have been sent to the Meteorological Service in the U.K. where they have been calibrated; they are now in the Deutsche Wetterdienst in Hamburg. This round robin trip will permit the determination of the spread of the values for some instruments according to the methodology used in each national radiometric centre.

In Cardiff some KIPP and Zonen pyranometers have been characterized versus the azimuth and cosine effect and a documentation on the photovoltaic cell pyranometers is being compiled.

III. RESULTS

The results of the comparison are summarized in the annex. They show it is possible to classify the pyranometers which participated in this task into four groups :

Group 1 : the Eppley PSP pyranometers

Group 2 : the old Kipp and Zonen type pyrameters such as G18, CM2, the old CM5, and the old Eppley multijunction pyranometers

Group 3 : the Kipp and Zonen CM5 pyranometers

Group 4 : the Kipp and Zonen CM10 pyranometers.

The groups were determined by the shape of the hourly values curves. By comparison with the number two reference, it seems that the curves which are the most horizontal, i.e. those which look most like a "perfect pyranometer" seem to be those of the third group. For the other groups, although the daily amount of measured radiation can be considered as similar, some care should be taken for the hourly values.

The calculation shows that the application for each group of pyranometers of a calibration factor instead of a calibration function for latitudes between 40 and 55 degrees shows a difference of about one or two per cent, according to the type of pyranometer, on the daily values if the calibration factor has been well chosen.

If the calibration factor at normal incidence is taken as a reference, the calibration factor used for field applications should be multiplied by 0.984 for group 1 pyranometers, 1.010 for group 2, and 0.997 for group 3.

The results and the proposed calibration factors are given in the annex.

The results of the pyranometer characterization established in the laboratory against a solar simulation test device show an appreciable difference between various pyranometers of the same CM5 type, but manufactured in different years. If the percentage deviation is very high, it would be interesting to recalculate it in terms of energy, which is more useful for solar energy applications.

The round robin of the two pyranometers is not yet finished but the calibration factors given by the French and British Meteorological Services came within 1 per cent of each other. The German Meteorological Service hopes to finish the calibration of these two instruments in the next few weeks.

IV. CONCLUSIONS

The results of the comparison bring out the necessity to coordinate the calibration procedures in the various countries of the EC area. It seems that the calibration factor of the substandard pyranometers of some countries needs to be changed. A new calibration campaign will be organized, employing the same method, at the beginning of 1982, with the same instruments.

If the results are confirmed, a methodology based on the full characterization of the substandard pyranometers will be proposed for the pyranometer calibration in 1982.

The first part of the booklet on the use of pyranometers has been written, it is planned to finish the draft before the end of 1981. After translation, it will be discussed by the working group on pyranometry in April 1982. At the same time the photocell pyranometers will be checked in laboratory to determine the accuracy obtainable in normal use.

Action 1 will issue :

- a methodology for the calibration of transfer and field pyranometers;

- a handbook on the use of pyranometers, giving users the main recommendations they must follow for accurate measurement.

ANNEX

Comparison of EC pyranometers and proposed calibration factors

Country	Group	Serial Number	Cal. factor before	Ratio Pyr/Ref 2	Group Adjustment	Proposed Cal. factor
Belgium	1	19682 F 3	98.9	0.965	0.974/0.984	94.4
	2	6217 A	118.6	1.008	1.013/1.010	119.9
Denmark	4	763327	131.8 at 15°C	0.948	0.995/0.997	124.9
Eire	2	690187	108.1 at 15°C	1.005	1.013/1.010	109.0
France	1	16542 F 3	92.5	0.991	0.974/0.984	90.8
	1	16540 F 3	93.8	0.992	0.974/0.984	92.1
	3	800082	56.5	0.988	-	55.8
	4	796617	116.4 at 15°C	0.996	0.995/0.997	115.7
	4	795546	110.2 at 15°C	1.010	0.995/0.997	111.1
F.R.G.	3	790057	57.5	1.012	-	58.2
Greece	1	14899 F 3	92.5	0.973	0.974/0.984	89.1
Netherlands	2	1564	113.2 at 15°C	1.005	1.013/1.010	114.0
U.K.	2	2508	122.5 at 15°C	1.003	1.013/1.010	123.2

ACTION 2

REFERENCE YEARS

Action leader : H. LUND

Thermal Insulation Laboratory, Technical University of Denmark, Lyngby, Denmark

Action 2 - SHORT REFERENCE YEARS

Action budget (CEC contribution) : 178.6 kUC

Duration of project : until 1983-06-30

Action Leader : H. LUND
Thermal Insulation Laboratory
Technical University of Denmark
Building 118
DK - 2800 LYNGBY, Denmark

Participants :

- Technical University of Denmark
  Contract Nr. ESF-011-DK(G)
  H. LUND, S. EIDORFF

- Delft University of Technology
  Contract Nr. ESF-010-80-N(B)
  A.W. BOEKE, A. VAN PAASSEN; H. LIEM

- Analysis and Development of Energy Systems
  Contract Nr. ESF-007-I(S)
  C. MUSTACCHI, V. CENA; F. HAGHIGHAT

- Technological Institute, Denmark
  Contract Nr. ESF-012-DK(G)
  L. HALLGREEN

- Institut Royal Météorologique de Belgique
  Contract Nr. ESF-003-B(G)
  R. DOGNIAUX, M. LEMOINE

Task :

Development and production of a library of data tapes; various sites inside the European Community.

0. Summary

The aim of the project is to develop 4 operational methods for generating Short Reference Years (SRY) from existing basic ideas or procedures. These methods will be evaluated, and the best method will be used for generation of SRYs for a number of locations within EEC.

Two of the methods generate SRYs from deterministic and stochastic properties extracted from the long term weather data set, one method by harmonic synthesis and another method by adjusting the distributions of the weather parameters in selected short periods in accordance with the distributions of the long term weather data set.

The evaluation will be done by comparing simulation runs with the different SRYs and the weather data set from which the SRYs have been derived.

One SRY generating method is ready, the other 3 methods and the common evaluating method will be ready by the end of 1981. The project is progressing according to the scientific plans. The recommended methodology for SRY will be ready mid 1982, and production of SRYs for a number of locations will be finished mid 1983.

Test Reference Years, TRYs, are now being produced or are ready for approx. 25 locations. For two countries raw data are still not available.

1. Introduction

Many computer programs using hourly weather data as Test Reference Years (TRYs) are rather costly to run, because of the long computing time. If a Synthetical Test Reference Year can be constructed, it should be possible with only small losses in resolution to generate it with considerably less than 365 days and still obtain values, distributions and relative frequencies like the full year, and realistic cross correlations. Such a data collection could be named a Short Reference Year (SRY).

Another method, which is also investigated in this project, is by careful selection of short runs of days to compile a SRY from real, measured weather data. These data are then adjusted in such a way that the distributions of the various parameters are in accordance with the original long term weather data set.

It is important that SRY must be representative for a wide variety of applications, because they will typically be used as input data for simulations, including both the building and the solar energy system.

The purposes for the project are to give the participants an opportunity to refine, generalize and finish their own methods for generations of SRY, by giving them access to weather data files from other locations in Europe, and finally, after a comparison between the methods and the SRY they produce, to give a recommendation regarding the methods. Another purpose will be to establish thumb rules for the minimum length of SRYs for various purposes.

As a by-product of the project SRYs will be generated for some locations in Europe.

2. Fundamentals of the candidate SRY generating methods

In the methods from Prof. Boeke, Mr. van Paassen and Mr. Liem the hourly values of the weather parameters are generated by means of deterministic daily variations modulated by daily and hourly random noise. The primary parameter is the amplitude Aq of a deterministic function giving the hourly variations of the global radiation. Aq is determined by a stochastic function with a probability density in accordance with the long term weather data set. According to Aq daily deterministic functions of the other weather parameters are selected and modified by stochastic functions with statistic properties, such as daily means, standard deviations, auto correlations, and cross correlations, according to the long term weather data set.

The method from Prof. Mustacchi and Mr. Cena is basically the same except for the extraction of the lower frequencies.

In the method from Mr. Hallgreen the hourly weather data are generated harmonically from a time series analysis of the long term weather data set. The lowest periods time is 8 hours. On the calculated radiation values a stochastic variation is superimposed.

In the method from Mr. Eidorff a SRY is compiled of short periods, several days, from the long term weather data set selected in such a way that they are typical for the period they are to represent. The original data in these short periods are then adjusted, as little as possible, so that the distributions of the various parameters are in accordance with the said periods in the long term weather data set. The best adjustment procedure has been determined by experiments with full length TRYs. The more data in the TRYs give a staticticically stronger discrimination between the different procedures. Table I shows an example of these tests. The adjustments of the distributions are based on hourly values of dry bulb temp., direct normal radiation and diffuse radiation on horizontal.

It is agreed in the group that all methods shall comprise a fully automatic procedure for complete long term weather data sets and be so well documented that special treatment of certain data sets could be designed.

3. Evaluation of the SRY generating methods

The evaluation of the candidate methods is based on simulation runs with solar energy systems and building thermal loads. The results of simulation runs on a certain system or building with the different SRYs are compared with a simulation run with the long term weather data set from which the SRYs are generated.

In order to select the best method on a high confidence level it is necessary to perform these simulations on an appreciable number of systems and buildings with weather data sets from several locations.

The simulation program for the solar energy systems is designed by Prof. Mustacchi. The program for the building thermal load simulations is BA4 with the necessary enhancements for use with SRYs designed by H. Lund.

The output of a certain simulation run shall be a parameter comparable with the output from the other simulation runs. The nature of this parameter is still under consideration, but it is agreed that it shall be in accordance with the normal field of application of SRYs.

Preliminary is proposed as such parameters heating and cooling loads for buildings and fuel consumption for domestic hot water systems. Yearly and seasonal values are considered as well as fractiles in the distributions of half-hourly and daily values.

The main comparisons will be carried out with SRYs consisting of 2 weeks per season. SRYs with one week per season and even shorter will be included in the evaluation.

The long term weather data sets will primarily be those from the earlier TRY evaluating project and some of the data sets from the TRY production program under action 2.

4. Status and preliminary results

The method from Prof. Boeke et al is in principle ready and able to generate SRYs from most data sets. Charts with statistical properties have been presented. Test runs have been made with Dutch weather data, and is now continuing with Belgian data.

The method from Prof. Mustacchi and Mr. Cena is ready to generate SRYs with radiation and temperature. Later wind and humidity will be added. Graphs with statistical properties have been presented. Test run has been made with data from the VIA1 tape (7 European locations) and de Bilt.

Mr. Hallgreen is not yet ready with the mathematical model. The method will be ready by the end of this year.

Mr. Eidorff will be ready with the selection step in his method by primo December. The adjustment step is ready, and tables showing the improvement in the simulation results obtained with an adjusted full year TRY compared to the original TRY have been presented (Table I).

The contents and some technical details of the SRYs have been decided. A list of possible users for SRYs and their fields of application have been worked out.

5. Conclusion regarding SRY work

The SRY project in action 2 is progressing according to the scientific plans and also largely according to the time table.

6. Production of Test Reference Years, TRY

Test Reference Years are now being produced by mr. Dogniaux according to the method recommended during the earlier project period. Tapes for 17 stations are ready, and 8-10 more are in production (Table II). Not all of the TRYs give the radiation split up into diffuse radiation and direct normal radiation, which is highly desirable for the users. As soon as data and methods are ready this splitting up must be done.

In some member countries there are still problems about finding a suitable distribution channel to the potential users.

Example: Yearly thermal loads at a constant room temp. of 22°C

Room	Weather data	Total cooling (Average per year, kWh)	Total heating (Average per year, kWh)
Heavy	15 years	706.5	2981.7
South	TRY Danish	-6.5	92.6
15% window	TRY adjusted	-2.9	3.6
Lightweight	15 years	2141.4	3357.7
South	TRY Danish	-39.3	-108.0
30% window	TRY adjusted	-5.2	8.5
Heavy	15 years	141.8	3672.3
North	TRY Danish	11.0	-136.9
15% window	TRY adjusted	-1.7	0.2

Table I. Example. Improvement of results through distribution adjustment. From TRY applications with one specific application program and for one specific "TRY adjustment procedure" as it will be tested for SRYs.

The figures for the 15 years period are average per year, kWh, for "TRY Danish" and for "TRY adjusted" are given the deviations from the 15 years figures.

PRESENT STATE OF COMPUTATION OF TRY NOVEMBER 19, 1981

COUNTRY STATION	SS	G	D	T	WS	U	PERIOD	REMARKS
BELGIUM								3 finished
UCCLE	H	H	H	H	H	H	58 - 75	
OOSTENDE	H	H		H	H	H	62 - 80	Try tape
SAINT-HUBERT	H	H		H	H	H	62 - 80	Available
DENMARK								1 finished
VÆRLØSE	DS	H	H	H	H	H	59 - 73	Try tape
FRANCE								6 finished
NANCY	DS	H		H	H	H	68 - 78	Try tape
TRAPPES	DS	H	H	H	H	H	68 - 78	Available
MACON	DS	H		H	H	H	68 - 78	
LIMOGES	DS	H		H	H	H	68 - 78	
CARPENTRAS	DS	H	H	H	H	H	68 - 78	
NICE	DS	H		H	H	H	68 - 78	
GERMANY								
NORDERNEY	H	H		H*	H*	H*	67 - 75	Months selected
HAMBURG	H	H	H	H	H	H	66 - 75	
WURZBURG	H	H		H	H	H	66 - 75	Tape in
HOHENPEISSENBERG	H	H		H*	H	H*	66 - 75	preparation
GREAT BRITAIN								4 finished
LONDON	H	H	H	H	H	H	59 - 68	Try tape
ESKDALEMUIR	H	H	H	H	H	H	59 - 68	available
ABERPORT	H	H	H	H	H	H	59 - 68	
LERWICK	H	H	H	H	H	H	59 - 68	
IRELAND								
VALENTIA		H		DM			69 - 79	Months selected
KILKENNY		H		DM			69 - 79	Hourly data expec-
DUBLIN	DS			DM			69 - 79	ted
CLONES	DS			DM			69 - 79	
BELMULLET	DC			DM			69 - 79	
MALIN HEAD	DS			DM				
THE NETHERLANDS								3 finished
DE BILT	H	H	H	H	H	H	71 - 80	Data received on
EELDE	H	H		H	H	H	71 - 80	November 8th
VLISSINGEN	H	H		H	H	H	71 - 80	
ITALY	DATA NOT YET AVAILABLE							
GREECE	DATA NOT YET AVAILABLE							

MEANING OF SYMBOLS: DS: daily sum — WS: wind speed
DM: daily mean — U : humidity
H : hourly value — T : temperature
Period: years from which the TRY is selected. — (*): except for some periods

Table II. Test Reference Years finished or in preparation, status Nov.19,81.

ACTION 3

ATLAS AND DATA BOOKS (FROM EXISTING DATA)

Action 3.1 - Global irradiance on horizontal planes - Definition of radiation climate zones in the EC
Action leader : F. KASTEN, Deutscher Wetterdienst, Meteorologisches Observatorium Hamburg, Federal Republic of Germany

Action 3.2 - Global irradiance on tilted planes - direct radiation
Action leader : R. DOGNIAUX, Institut Royal Météorologique de Belgique, Brussels, Belgium

Action 3.3 - Statistical analysis of meteorological data
Action leader : J.A. BEDEL, Direction de la Météorologie, Service Météorologique Métropolitain, Paris, France

Action 3.4 - Sensitivity analysis of useful energy output for solar converters with respect to quality and completeness of meteorological data sets
(paper prepared and presented by H.R. KOCH, Kernforschungsanlage Jülich, Federal Republic of Germany)

Action 3.1 : GLOBAL IRRADIANCE ON HORIZONTAL PLANE DEFINITION OF RADIATION CLIMATE ZONES IN THE EC

Action Leader : F. KASTEN
Deutscher Wetterdienst
Meteorologisches Observatorium Hamburg
Frahmredder 95
D - 2000 HAMBURG 65

Participants :

- Deutscher Wetterdienst
  Contract no. ESF-004-80-D(B)

- Institut Royal Météorologique de Belgique
  Contract no. ESF-003-B(G)

Tasks :

Improvement and extension of the first volume of the EC solar radiation atlas published in 1980

Definition of radiation climate zones in the EC

Summary: For improving the details of the existing EC solar radiation atlas, additional radiometric and particularly heliographic stations have been taken into account; processing of these data has advanced to the phase of producing the final tables. In order to extend the area covered by the existing atlas, the radiation data of many adjacent countries have been retrieved from several sources and have reached different stages of processing, some of them being ready for final production. Special treatment was given to the Mediterranean area where the analysis had mostly to rely on sunshine duration data.

1. Introduction

The first volume of the European Solar Radiation Atlas published by the Commission of the European Communities (1) presents preliminary monthly maps of mean daily global irradiation on the horizontal plane. The isolines of global irradiation also called isopyrs were designed on the basis of the 10 years (1966-1975) data recorded at 56 radiometric stations within the EC.

The monthly maps shall be improved with respect to both more details of the isopyrs in the region of the EC and better extrapolation of the isopyrs across the borders of the EC.

2. Filling the gaps in the existing atlas

In order to improve the details of the isopyrs in the region of the EC, the gaps between the 56 radiometric stations mentioned above have to be filled. In the absence of additional radiometric stations, the daily sunshine duration data of heliographic stations located in between the radiometric stations are to be used.

The radiometric networks in the Netherlands and in Belgium were felt to be sufficiently dense for the present purpose so that additional data from these two countries were not required.

From Denmark, global radiation tables of the following 3 stations outside the Danish meteorological network have been acquired, punched on cards and quality-controlled:

Stevns
Karup
Jyndevad

The Italian meteorological service provided global radiation data on tape of the following additional 15 stations:

Udine	Pisa	Crotone
Torino	Viona de Valle	Ustica
Venezia	Pescara	Gela
Bologna	Amendola	Olbia
Capo Mele	Capo Palinuro	Cagliari

These data have been re-formatted, quality-controlled and then computer-printed on tables of the same format as in the existing atlas. An example is given on Table 1.

From Ireland, the United Kingdom, France and the Federal Republic of Germany, data of the heliographic networks were taken into account. Daily sunshine duration data of the following numbers of heliographic stations were provided on tape by the meteorological services:

Ireland	14
United Kingdom	38
France	44
Germany (F.R.)	68

These heliographic stations are to be associated as "satellites" to the appropriate radiometric stations which are to be considered as representative for the radiation climate in the respective areas. Tab. 2 shows an example of this association of heliographic to radiometric stations as suggested by the French meteorological service.

From all radiometric stations in the EC listed in the existing atlas, the daily sums of global radiation G and sunshine duration S of the period 1966–1975 were used to establish the Ångström relation

$$G/G_0 = a + b \cdot S/S_0 \qquad (1)$$

for each month and each station. G_0 and S_0 are the corresponding values of extraterrestrial solar radiation and astronomical sunshine duration, respectively. The regression coefficients a and b were determined by the least squares method from the approximately 300 pairs of G/G_0 and S/S_0 per month and per station.

The regression coefficients a and b as well as the corresponding correlation coefficients R were summarized on tables. As an example, these constants for the 7 French radiometric stations listed in the existing atlas are reproduced on Table 3.

With the regression coefficients a and b of the radiometric stations, the monthly means of daily sunshine duration S of the satellite heliographic stations are converted to monthly means of daily global irradiation G, for each of the 120 individual months of the 10 year period 1966–1975; then, the 10 year means of the monthly means are computed. This task has been completed for Germany and France.

Finally, the values are entered on geographical maps and interpolated by drawing isolines. The latter has to take due account of the geographic and climatic conditions of the respective area and shall therefore be based on the advice requested from the experts of the national meteorological services. In the case of Germany, this task has been completed and the results have recently been published in the literature (2).

3. Extending the area of the existing atlas

For a better extrapolation of the isopyrs across the borders of the EC, data of daily global irradiation of the period 1966–1975 were requested from several European countries.

Yugoslavia sent a tape with the complete data of the following 11 stations:

Ljubljana	Zlatibor
Zagreb	Split
Banja Luka	Pristina
Beograd	Bar
Negotin	Bitola
Sarajewo	

Austria provided the complete data partly on tape, partly on tables which were transferred on punched cards, of the following 5 stations:

Innsbruck	Sonnblick
Klagenfurt	Wien
Salzburg	

Complete tables of daily G and S of 4 stations were received from Poland. These data were transferred on punched cards, quality-controlled, formatted and are ready for printout as atlas-tables. The 4 stations are:

Kolobrzeg	Warszawa
Suwalki	Zakopane

Switzerland sent data tables of 4 stations 3 of which are still incomplete at the present but will be completed in the near future:

Davos	Zürich
Weissfluhjoch	Locarno

The complete data of Locarno were punched and quality-controlled.

Of 6 radiometric stations in Spain data were available from, only Madrid has a complete series of data being useful for the atlas. These data have been quality-controlled.

The meteorological service of Norway provided tables of daily G and S of 6 stations but only for parts of the desired period 1966-1975. A tape with a complete data set was received from the University of Bergen.

Data were not received so far from Sweden, Portugal, Czechoslovakia and the German Democratic Republic. Of the latter, daily global irradiation data of the 4 stations

Heiligendamm	Dresden
Potsdam	Fichtelberg

were taken from the publications of the WMO World Radiation Data Center at Leningrad, transferred on punched cards and quality controlled.

4. Special treatment of the Mediterranean area

Data of daily global irradiation and predominantly of daily sunshine duration were collected, among others, from Marocco, Tunesia, Algeria, Malta, Egypt, Israel, Lebanon, Jordan, Syria, Cyprus, Turkey, Romania and Greece. On the basis of the sunshine data, maps of mean monthly and yearly sunshine duration, in hours, were drawn. An example is shown in Fig. 1.

By graphical interpolation on these maps, the mean sunshine duration at the grid points of a 1° by 1° latitude and longitude grid ranging from 44°N to 30°N and from 16°E to 46°E is determined.

These sunshine duration data will be associated to the global irradiation data of the radiometric stations being available in the Mediterranean area. Then, linear regression on the basis of the Ångström formula as described in section 2 will be applied to the data in order to convert daily sunshine duration to daily global irradiation. Finally, maps of mean daily global irradiation of the Mediterranean area will be drawn.

5. Conclusions

The subtask "Filling the gaps of the existing atlas" has progressed to the production phase. The computer-printing of final data tables for the atlas is under way with the exception of Ireland and the United Kingdom which countries are requested for advice on associating the heliographic stations to the appropriate radiometric stations.

In the subtask "Extending the area of the existing atlas", the phase of data collection, processing, re-formatting and quality control has reached different stages for the different countries. In a few cases, production can be started immediately. From some countries, the data of only 1 station are suitable for treatment for the present purpose. The data of a few countries are to be taken from the publications of the WMO World Radiation Data Center for lack of other data sources.

With regard to the subtask "Mediterranean area", appreciable progress in data collection and processing has been made. Special efforts will be made to integrate the available data from Greece into the EC atlas.

Literature

(1) Commission of the European Communities: European Solar Radiation Atlas, Vol I: Global Radiation on Horizontal Surfaces, ed. by W. Palz. Grösschen-Verlag, Dortmund (1979).

(2) H.J. Golchert: Mittlere monatliche Globalstrahlungsverteilungen in der Bundesrepublik Deutschland. Meteorol. Rundschau 34, 143-151 (1981).

ITALIA STATION VENEZIA TESSERA

LATITUDE 45 30'N LONGITUDE 10 20'E ALTITUDE 6 M

GLOBAL RADIATION G IN WH M-2 (WRR) SUNSHINE DURATION S IN 1/10 H

10 YEAR MEANS (1966-1975) OF MONTHLY MEANS OF DAILY SUMS

	JAN	FEB	MAR	APR	MAY	JUN	JUL	AUG	SEP	OCT	NOV	DEC	YEAR
G	1121	1897	3518	4690	5809	6640	6989	5895	4341	2855	1416	1104	3878
GMAX	2198	3639	5709	7246	8053	8691	8731	7825	6208	4577	2689	1980	5655
GMIN	195	293	648	829	1572	1948	2882	2014	1102	675	242	250	1063
GO	3344	4909	7042	9265	10907	11616	11236	9857	7829	5590	3751	2902	7386
G/GO	.34	.39	.50	.51	.53	.57	.62	.60	.55	.51	.38	.38	.49
S	*****	*****	*****	*****	*****	*****	*****	*****	*****	*****	*****	*****	*****
SMAX	*****	*****	*****	*****	*****	*****	*****	*****	*****	*****	*****	*****	*****
SO	90	102	118	134	147	154	151	139	124	108	94	86	121
S/SO	*****	*****	*****	*****	*****	*****	*****	*****	*****	*****	*****	*****	*****

Table 1 : Example of the final tables for the atlas of the 10 year means (1966-1975) of monthly means of daily sums of global radiation G and sunshine duration S. Venezia; Italia

TRAPPES

LILLE
ABBEVILLE
BEAUVAIS
CAEN
RENNES
LE BOURGET
ORLEANS
AUXERRE
LE MANS
TOURS
BOURGES
NANTES
BREST
ILE DE BATS
ILE DE BREHAT

NANCY

REIMS
METZ
STRASBOURG
COLMAR

MILLAU

LE PUY

NICE

TOULON
AJACCIO
BASTIA

CARPENTRAS

MONTELIMAR
EMBRUN
MARIGNANE
NIMES
MONTPELLIER
PERPIGNAN

LIMOGES

POITIERS
LA ROCHELLE
BORDEAUX
CAZAUX
AGEN
MONT DE MARSAN
PAU
TOULOUSE

MACON

DIJON
BESANCON
NEVERS
VICHY
CLERMONT-FERRAND
LYON
AMBERRIEU

TAB. 2: 44 HELIOGRAPHIC STATIONS ASSOCIATED WITH 7 RADIOMETRIC STATIONS IN FRANCE (AS SUGGESTED BY MÉTÉOROLOGIE NATIONALE).

STATION	JAN	FEB	MAR	APR	MAY	JUN	JUL	AUG	SEP	OCT	NOV	DEC
						a						
PARIS/TRAPPES	.18	.19	.20	.22	.22	.23	.24	.23	.22	.21	.19	.18
NANCY	.19	.20	.21	.21	.22	.21	.22	.23	.22	.19	.18	.18
MACON	.18	.19	.20	.19	.19	.21	.21	.21	.20	.19	.17	.17
LIMOGES	.19	.19	.20	.21	.22	.23	.21	.21	.21	.21	.19	.19
MILLAU	.19	.19	.20	.20	.19	.22	.23	.22	.21	.20	.19	.20
CARPENTRAS	.19	.17	.19	.20	.22	.23	.27	.23	.21	.20	.21	.20
NICE	.18	.18	.20	.21	.21	.23	.23	.22	.20	.21	.19	.20
						b						
PARIS/TRAPPES	.59	.54	.55	.53	.54	.52	.51	.49	.51	.51	.54	.57
NANCY	.60	.57	.57	.55	.57	.56	.54	.51	.53	.57	.58	.55
MACON	.61	.55	.55	.57	.57	.55	.53	.53	.55	.56	.58	.59
LIMOGES	.62	.60	.61	.58	.58	.55	.56	.55	.56	.54	.56	.57
MILLAU	.63	.62	.63	.63	.64	.59	.56	.57	.60	.60	.61	.59
CARPENTRAS	.50	.55	.58	.58	.56	.54	.48	.52	.55	.52	.48	.48
NICE	.55	.57	.57	.56	.56	.53	.50	.51	.54	.52	.54	.52
						R						
PARIS/TRAPPES	.92	.93	.95	.94	.94	.94	.95	.94	.95	.94	.93	.90
NANCY	.91	.93	.95	.95	.96	.96	.96	.96	.95	.95	.92	.86
MACON	.92	.96	.96	.97	.96	.96	.96	.96	.96	.95	.94	.93
LIMOGES	.93	.95	.96	.96	.96	.93	.97	.97	.96	.96	.94	.93
MILLAU	.94	.95	.96	.96	.96	.96	.95	.96	.97	.97	.96	.93
CARPENTRAS	.95	.96	.96	.97	.97	.96	.95	.96	.96	.96	.96	.95
NICE	.95	.96	.96	.96	.96	.95	.94	.96	.96	.94	.96	.94

TAB.3 : REGRESSION COEFFICIENTS a, b AND CORRELATION COEFFICIENT R OF THE REGRESSION $G/G_0 = a + b \cdot S/S_0$ FOR EACH MONTH, BASED ON THE DAILY SUMS OF GLOBAL RADIATION G AND SUNSHINE DURATION S, 1966–1975, FRANCE.

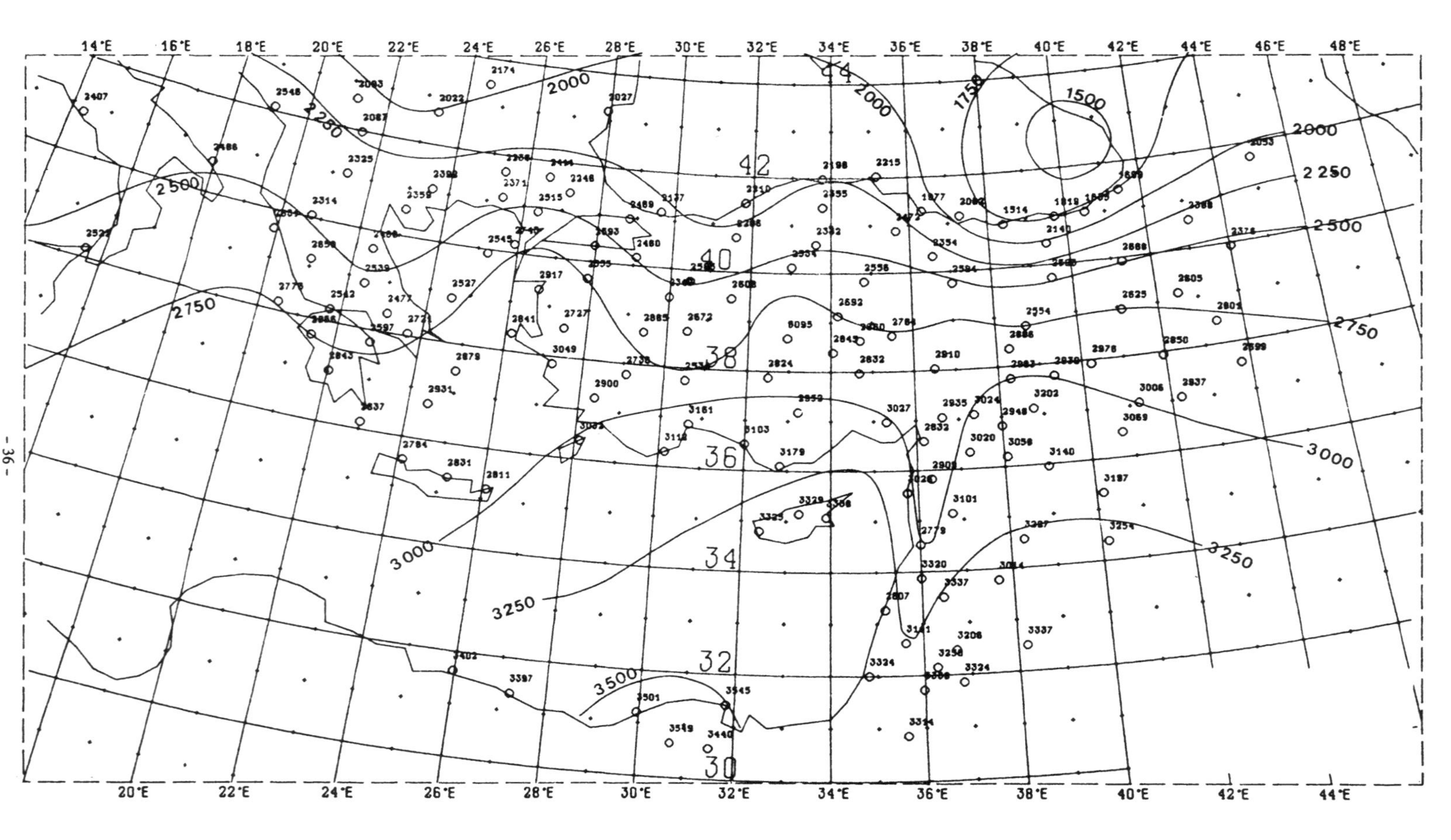

Fig. 1: Mean yearly sunshine duration, in hours, in the eastern Mediterranean area.

Action 3.2 - GLOBAL IRRADIANCE ON TILTED PLANES - DIRECT RADIATION

TASKS : GLOBAL IRRADIATION ON TILTED PLANES
DIRECT IRRADIATION ON HORIZONTAL AND TILTED PLANES
CALCULATION OF TURBIDITY FACTOR

ACTION BUDGET (EC CONTRIBUTION ONLY) : 156 kUC

ACTION LEADER : R. DOGNIAUX
INSTITUT ROYAL METEOROLOGIQUE DE BELGIQUE
AVENUE CIRCULAIRE, 3
B - 1180 BRUSSELS

PARTICIPANTS :

- INSTITUT ROYAL METEOROLOGIQUE DE BELGIQUE (CONTRAT N°ESF-003-B(G) : R. DOGNIAUX)
- METEOROLOGISCHES OBSERVATORIUM HAMBURG (CONTRAT N° ESF-004-D(B) : F. KASTEN)
- METEOROLOGIE NATIONALE- PARIS (CONTRAT N° ESF-005-F(G) : J.A. BEDEL)
- KONINKLIJK METEOROLOGISCH INSTITUUT (CONTRAT N° ESF-006-NL (B) : W.H. SLOB)
- CNRS LABORATOIRE D'ENERGETIQUE SOLAIRE (CONTRAT N° ESF-009-F(G) : J.F. TRICAUD)
- IRISH METEOROLOGICAL SERVICE (CONTRAT N°ESF-018-EIR (G) : E.J. MURPHY)
- LMT LICHTMESSTECHNIK GmbH - Berlin(CONTRATS N°ESD-290-77
ESF -020-D (D) : J. KROCHMANN)
- DEPT. OF BUILDING SCIENCE - UNIVERSITY OF SHEFFIELD (CONTRAT N° ESF-021-UK (H) : J.K.PAGE)
- METEOROLOGICAL OFFICE - U.K. (CONTRAT N° ESF-024-80-UK(H) : R. RAWLINS)

1. TASKS, METHODOLOGY AND ESTIMATED TIME ALLOWED

1,1. SUMMARY OF THE TASKS

- Calculation of global irradiation on vertical planes oriented South, East and West and on planes oriented southward at the angle of the latitude on the base of monthly means of daily values and, at a later stage, of hourly values.
- Calculation of the direct component of global irradiation on horizontal plane (tilted planes at a later stage) on the base of hourly values.
- Calculation of the turbidity factor.

1,2. METHODOLOGY

Title	Organisation
Improvement of simple calculation methods for inclined surfaces (action leader)	Institut Royal Météorologique de Belgique (B)
Parameterization of radiation fluxes as function of solar elevation, cloudiness and turbidity	Meteorologisches Observatorium Hamburg (D)
Calculation of global irradiance on inclined surfaces	Koninklijk Nederlands Météorologisch Instituut (NL)
Statistical analysis of measurements of radiation on inclined surfaces	Irish Meteorological Service (IR)
Climatological study of the diffuse radiation on tilted surfaces	Météorologie Nationale, Paris (F)
Estimation of hourly radiation over the U.K.	Meteorological Office, Bracknell (GB)
Production of data to prepare an atlas of radiation on inclined surfaces	Department of Building Science, University of Sheffield (GB)
Ditto (combined with Sheffield); angular distribution of sky radiance and ground reflected radiation fluxes	LMT LICHTMESSTECKNIK GmbH Berlin (D)
Statistical analysis of solar radiation data on differently oriented planes at a hight altitude station	CNRS/Laboratoire d'Energétique Solaire Odeillo (F)

1.3. PRODUCTS AND ESTIMATED TIME ALLOWED

1.3.1. Global irradiation on tilted planes

- Detailed comparison of the Page and Krochmann's methods and analysis of the outputs using selected data of the EEC solar radiation data tapes (July 1981)
- Production of a recommended method, finalisation of a recommended EEC radiation model and preparation of a user guide manual for a pratical computation based on the choice of suitable parametric quantities (December 1981)
- Production of data for tilted planes for extended number of stations (March 1982)
- Plotting and drawing of the curves isopyrs for the volume 2 of the atlas (May 1982)

1.3.2. Direct irradiation

- Production of data for the volume 3 of the atlas (June 1983)
- Atlas ready for printing (June 1983)

1.3.3. Other tasks

- Production of the final results : June 1983

2. PRESENTATION OF THE WORK SO FAR PERFOMED

N.B. For the convenience of the statement, the order of presentation will follow the scheme adopted in the previous action leader's documents prepared for the first and second contractors coordination meetings in Jülich 5th and 6th May, 1980 and in Brussels 1st and 2th April 1981.
For the easiness of the lecture, partial results and technical notes are given in appendix. Corresponding references are given into brackets in the text.

ITEM 1 : APPLICATION OF THE SELECTED METHODS FOR THE EVALUATION OF IRRADIATION ON TILTED PLANES; PRODUCTION OF DATA FOR THE ATLAS (UNIVERSITY SHEFFIELD (U.K.)/LICHTMESSTECHNIK (D)

The first part of tho project has been concerned with the detailed comparison and analysis of the outputs achieved from the two developed calculation methods, Page and Krochmann. For this purpose , comparison calculations have been carried out using selected data from the EEC solar radiation data tapes[1] ,The work was divided into the following phases : :

Phase 1 : 1 st April 1980 - 31 st August, 1981

Review of the remaining problems with other inclined surface specialists in the EEC, formalisation of algorithms for standard calculations, standardisation of definitions of concepts like sunrise, declination, etc.

Phase 2 : 1 st September, 1980 - 30th November, 1981

Verification of clear and overcast sky models [2] [3].

University of Sheffield - lead responsibility
Modify Sheffield system to Linke turbidity system and then check out of clear sky data for slopes using EEC existing data tapes and the Linke turbidity coefficient model and compare with Krochmann results. Crosschecks with the Berlin clear sky model.
Production, in liaison with University of Berlin, of a recommended EEC method for estimation of solar radiation on slopes on clear days.

Lichtmesstecknik Berlin - lead responsibility
Check out of overcast sky data using EEC data tapes.
Production of a recommended method in collaboration with the University of Sheffield for estimation of solar radiation on slopes on overcast days.

Phase 3 : 1 st December, 1980 - 31st January, 1981

Development of final average day model

University of Sheffield - lead responsibility
Assembly of data on average diffuse radiation on horizontal surfaces on a European basis; development of regression equations for average diffuse radiation prediction.

Lichtmesstecknik Berlin - lead responsility
Assembly of data on ground albedo characteristics of different parts of Europe.

University of Sheffield / Lichtmesstecknik Berlin - joint responsibility. Inter-comparison of Sheffield and Berlin models
Finalisation of recommended EEC radiation model for predicting hourly mean irradiation of slopes.
Verification of final model for a range of European stations observing solar radiation on slopes using EEC data tape.
Production of systematic tables of recommended values of turbidity for different types of site in different parts of Europe.

Phase 4 : 1 st February - 31st March 1982

Preparation of data for slopes for specific stations and preparation of user guide manual :

University of Sheffield
Preparation of a guide manual to a practical computation giving specific guidance on how to use the model and the choice of suitable parametric quantities, for example turbidity, in different parts of the EEC in different types of situation, towns, rural areas, etc.

Lichtmesstecknik Berlin
Using EEC data for the specific stations listed in the horizontal surface atlas, preparation of standard data sheets to an agreed format, using final EEC model developed in phase 3. The outputs would be a set of tables, station by station.

Phase 5 : 1 st April 1982 - 31 st May 1982

Production of the curves isopyrs for the inclined surface atlas related to the actual EEC horizontal surface atlas.

I.R.M. :
Drawing of the curves for the vertical planes oriented South, East and eventually West and on planes oriented southward at the angle of the latitude (monthly means of daily values).

Phase 6 : (in conjonction with phase 5)

The production of data and maps for a next volume of inclined surface atlas related to extended countries of Europe could not start until the corresponding EEC horizontal surface atlas is available. This phase of the project would be concerned with the detailed development of the final methodology for the production of the relevant data.
Note that the contract does not include provision for the actual production of the atlas which would require substantial additional sums of money for its completion. This completion phase could only take place after the revised horizontal surface atlas became available.

ITEM 2 : IMPROVEMENT OF THE METHOD OF CALCULATION OF IRRADIATION AN A HOURLY BASIS INCLUDING THE ESTIMATION OF THE DIFFERENT COMPONENTS. CORRELATION OF THE METHODS TO CLIMATIC ZONES.

Besides the studies reported in the preceding item, several methods for the separation of the direct and diffuse components of irradiation have been validated by another ways (KNMI-NL; MET OFFICE-UK; MET NATIONALE -F).

2.1 Its appears that the clearness index K_T gives comparable results as a measured sunshine duration [4]. For the calculations of the global irradiance on an inclined surface, different sky conditions are considered. Several models for an overcast, a clear and a partly clouded sky are validated. As a result of this study, it appears that the yearly global irradiance on the inclined surface can be calculated within a few percents if the most suited relations are used [5]. However the hourly standard deviation is in the order of 10-20% dependant on the available climatological parameters (KNMI).

2.2 The possibility of using the Ångström relation to estimate global, diffuse and direct irradiations averaged over a month or less from sunshine data is to be investigated.

The initial aim is to provide corresponding maps of hourly values of global and diffuse irradiations on horizontal surface over the UK, which has numerous sunshine recording stations and to provide, in a second stage, data of irradiations on inclined surfaces. The application of these schemes to other EEC countries with adequate sunshine records would also be studied.

The proposed method is dependent on the validity of the Ångström relation which has well known deficiencies and the accuracy is limited by the estimation of the daily coefficients for which errors arise both from their uncertainty at a particular site and from their spatial interpolation.

A wide range in global irradiations is found for months which are mostly overcast and also for hours near sunrise/sunset. This method yields relatively little information for these cases and an alternative empirical treatment may be necessary. Data for the number of completely overcast days in each month is not as readily available as sunshine totals; also fewer stations report hourly values.

The application of sunshine measurements for estimating the hourly diffuse irradiations is less useful than for global and measurements of diffuse irrradiation are made at fewer locations which will make the construction of maps more difficult (MET. OFFICE).

2.3 In view to improve the estimation of the diffuse component in the models of calculation of global irradiations on tilted surfaces, the MET-NAT-F has carried on the study of the diffuse radiation for clear sky irradiations by an analysis of the hourly sums of the diffuse radiation on different receiving surfaces.

This analysis shows that the spliting up of the diffuse radiation into two components (an isotropic one and a "direct" one)is not allowable. It is necessary to introduce a third component (a zenithal one). The latter can be either positive or negative. It is negative, in particular, for high values of the atmospheric turbidity (by clear sky condition) [6].

For the estimation of the diffuse radiation, it is suggested to use two parameters : the solar elevation and the "extented turbidity factor". The last parameter can be obtained by using the hourly values of direct solar radiation without taking account of the sunshine duration. It is calculated only when the direct solar radiation is over 1 J/cm^2. The three components of the diffuse radiation have been calculated for several values of sun's elevation.

ITEM 3 : PARAMETRIZATION OF RADIATION FLUXES AS FUNCTION OF OTHER METEOROLOGICAL PARAMETERS SUCH AS SUN'S ELEVATION, CLOUDINESS, TURBIDITY, ALBEDO AND RECEIVING SURFACE.

3.1 Systematic studies using the material collected from countries for which the extension of the atlas is anticipated are in progress for correlating the components of solar radiation with the relative duration of sunshine and for determining their variations on a geographical base as function of the radiation climatic area (IRM-B).

3.2 The equipment necessary to carry out the monitoring programme of measurements of the 3 components of solar radiation at Valentia Observatory(IR) in a vertical south-facing surface was delivered during the last week of March 1981. As all the preliminary preparations had been made, it was possible to install the instruments to measure the separate components of global radiation on a vertical south-facing surface without further delay. Testing and calibration of the instruments was carried out during April and data collection commenced on 1st May. Data collection and sorting has continued since then.

3.3 In the frame of the investigation of the dependence of solar and terrestrial radiation flux densities on cloud amount and cloud type incorporating turbidity as a kind of a second order effect conducted in the Met. Obs. Hamburg, (D) the compilation, completion and quality control of 1-hourly basic and 3-hourly supplementary cloud data of the 12 years time period 1964-1975 have been completed for 7 of 8 stations. For 4 stations, sets of tables and diagrams of mean hourly global irradiance as function of total cloud amount and solar elevation have been established [7]. From the data at Hamburg of mean hourly irradiances of global radiation and diffuse radiation at cloudless sky, the Linke turbidity factor has been determined and corresponding diagrams have been established for different classes of elevation of the sun.

3.4 With respect to the analysis of the solar radiation data on a high altitude station, in particular the distribution of diffuse component for different conditions of turbidity of duration of sunshine and of environment considered by the Lab. d'Energétique Solaire (F), the equipments required for the records of data have been received and after some problems for the starting up of the data logger, the measurements started in July. In the same time, data for the years 1979, 1980 and the beginning of 1981 have been put on punch cards.

ITEM 4 : STUDY OF THE ANGULAR DISTRIBUTION OF SKY RADIANCE AND GROUND REFLECTED RADIATION FLUXES (Swiss Meteorological Institute, Zürich)

Using the complete equipment of the mobile system of the Swiss Meteorological Institute which now also includes the 10 m mast with a pyranometer and a sensor of 5 degrees full view angle, 650 new series of measurements have been made at two different locations; the evaluation of data is in course.

The analyses of the angular dependence of radiance distribution over the upper as well as the lower hemisphere expressed as ratio of zenith radiance are made and studied.

The ratios of diffuse irradiance of tilted surface to that of the horizontal have been plotted against azimuth, tilt angle, solar height and turbidity already for a great number of cloudfree cases. The multiparameter-relationship starts to get definite patterns.

The digitalization of fish-eye photographs in order to parametrize amount, type and distribution of clouds over the sky has been settled for the whole material till the end of 1980. The evaluation of the new 650 pictures is in course.

CONCLUSION

Most of the individual tasks involved in the Action 3.2 programme unfolds in a orderly manner according the foreseen timing.
Some of the tasks which do not present a character of priority opposite one another were postponed without prejudice for the finality of the Action.
Two action meetings are foreseen in 1982, the first one in London, in January.

APPENDIX

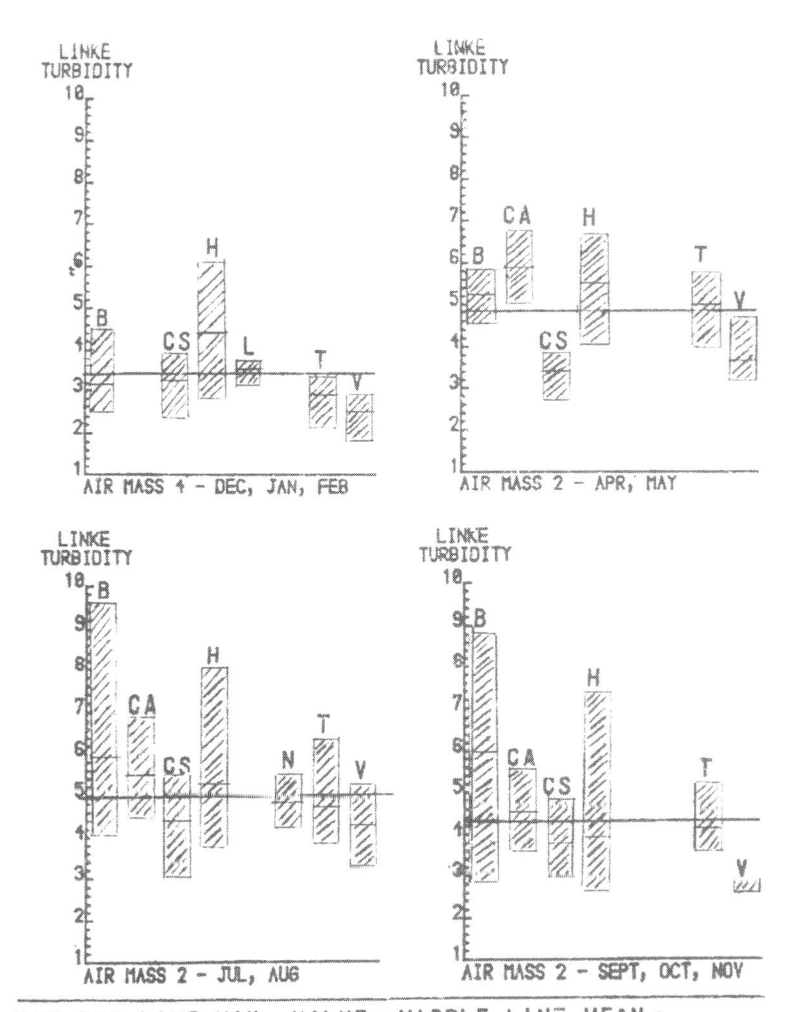

KEY: TOP LINE MAX. VALUE : MIDDLE LINE MEAN :
BOTTOM LINE MIN.VALUE, FOR STATIONS AS FOLLOWS

B BRACKNELL	H HAMBURG	T TRAPPES
CA CABAU	L LOCARNO MONTI	V VALENTIA
CS CARPENTRAS	N NORRKOPPING	

[1] Check - out of the dependance of the Linke turbidity factor on the air mass for clear sky conditions data for different months and different stations.
[University of Sheffield]

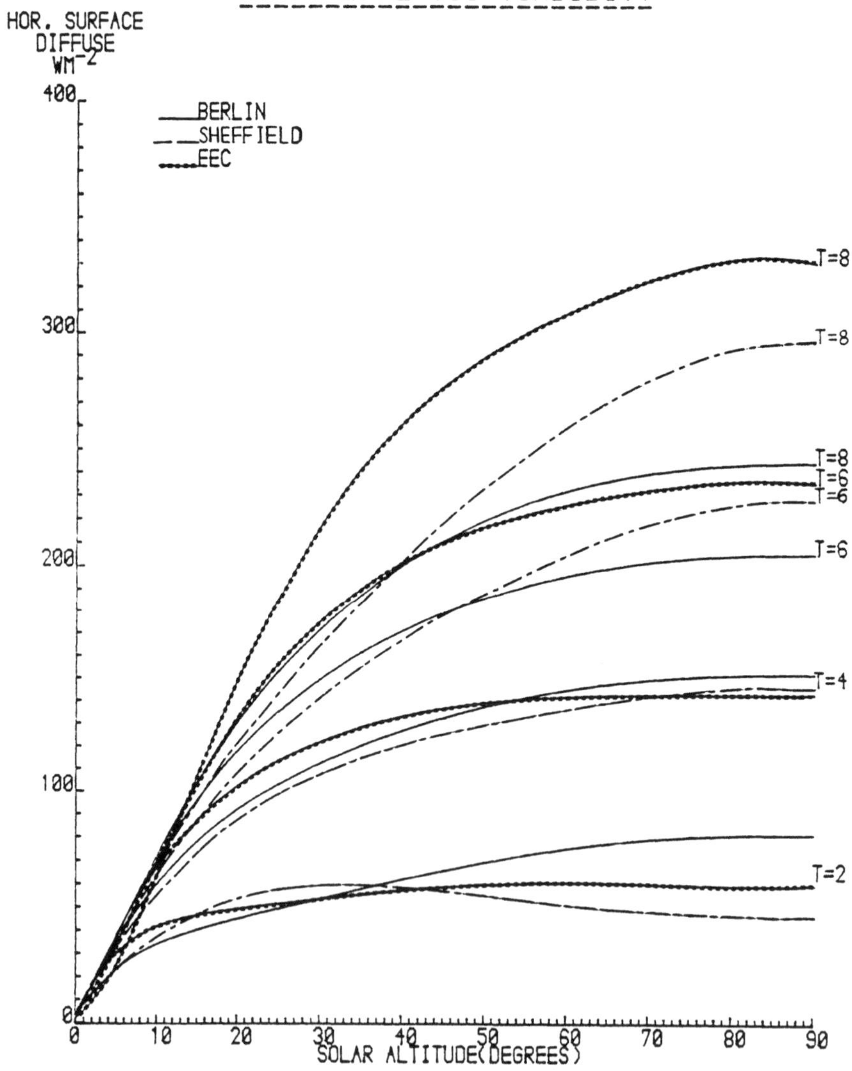

[2] Two horizontal diffuse models were checked for the development of appropriate regression equations for diffuse radiation prediction,
[University of Sheffield]

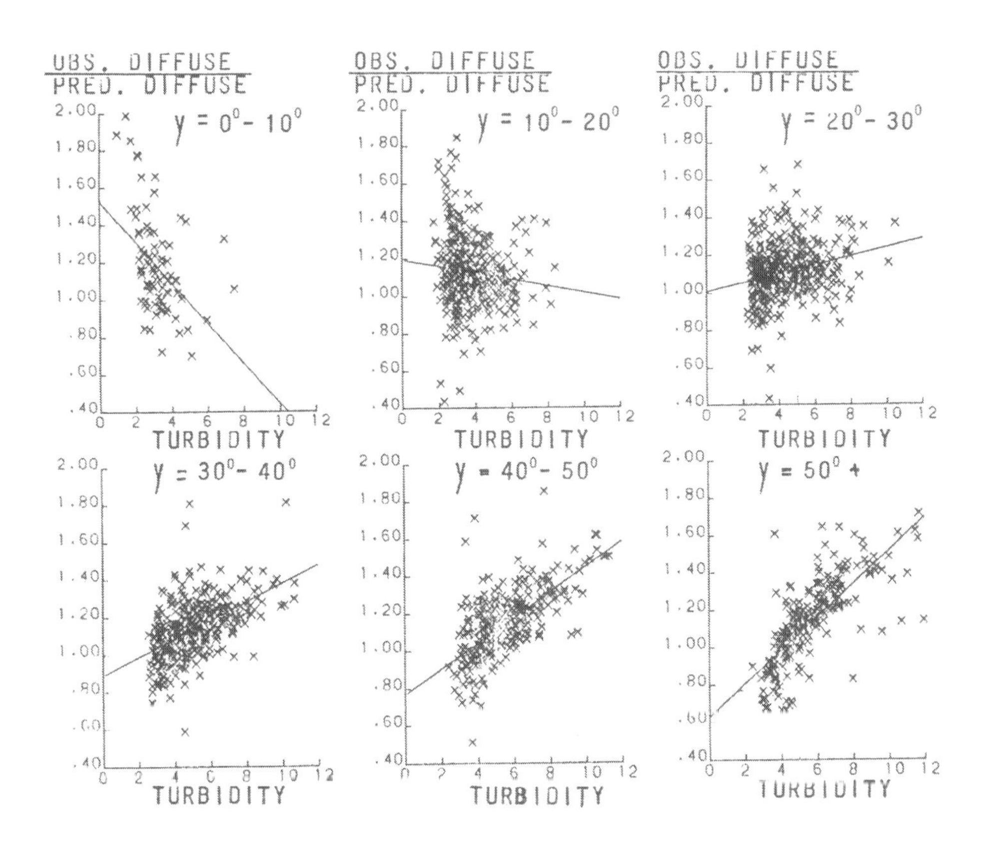

[3] Check - out of Berlin diffuse horizontal surface model by solar altitude class - all stations showing regressions between modelling errors and Linke turbidity factor at different solar altitudes.

Relations : $K_d = a K_T + b \sin (c K_T)$

$K_d = Gd/G_o$, $K_T = G/G_o$

Reference	a	b	c(rod)
Liu & Jordan	0.294	0.1445	4.97
Ruth & Chant	0.206	0.215	3.93
Bruno	0.31	0.139	4.62

Orgill & Hollands :

$K = Kd/K_T = Gd/G$

$K = 1. -0.249 K_T \quad 0<K_T < 0.35$

$K = 1.557-1.84 K_T \quad 0.35<K_T < 0.75$

$K = 0.1777 \quad 0.75<K_T$

Reference	Ratio of calculated and measured diffuse irradiance
Liu & Jordan	0.84 ± 0.19
Ruth & Chant	0.98 ± 0.21
Bruno	0.87 ± 0.19
Orgill & Hollands	0.99 ± 0.22
Franszen	1.01 ± 0.20

[4] Comparison of several empirical relations for the separation of direct and idffuse irradiance on a yearly basis with hourly standard deviations. [KNMI]

Available climatological parameters RATIO (standard deviation)					
Orientation		G	G, s/s_o	G,Gd	G,Gd, s/s_o
East	90	1.02(0.16)	1.03(0.19)	1.04(0.14)	1.05(0.14)
South	90	1.02(0.16)	1.01(0.20)	1.01(0.12)	1.01(0.11)
South	67^5	1.01(0.14)	1.02(0.16)	1.00(0.10)	1.01(0.09)
South	45	1.01(0.13)	1.02(0.15)	1.00(0.09)	1.02(0.10)
South	22^5	0.99(0.10)	1.00(0.10)	0.98(0.07)	0.99(0.07)
West	90	1.02(0.16)	1.01(0.14)	1.01(0.12)	1.02(0.11)
North	90	0.98(0.12)	1.02(0.13)	1.01(0.14)	1.03(0.13)

[4] RATIO of the calculated irradiance and the measured on an inclined surface for hourly values over a period of one year.
[KNMI]

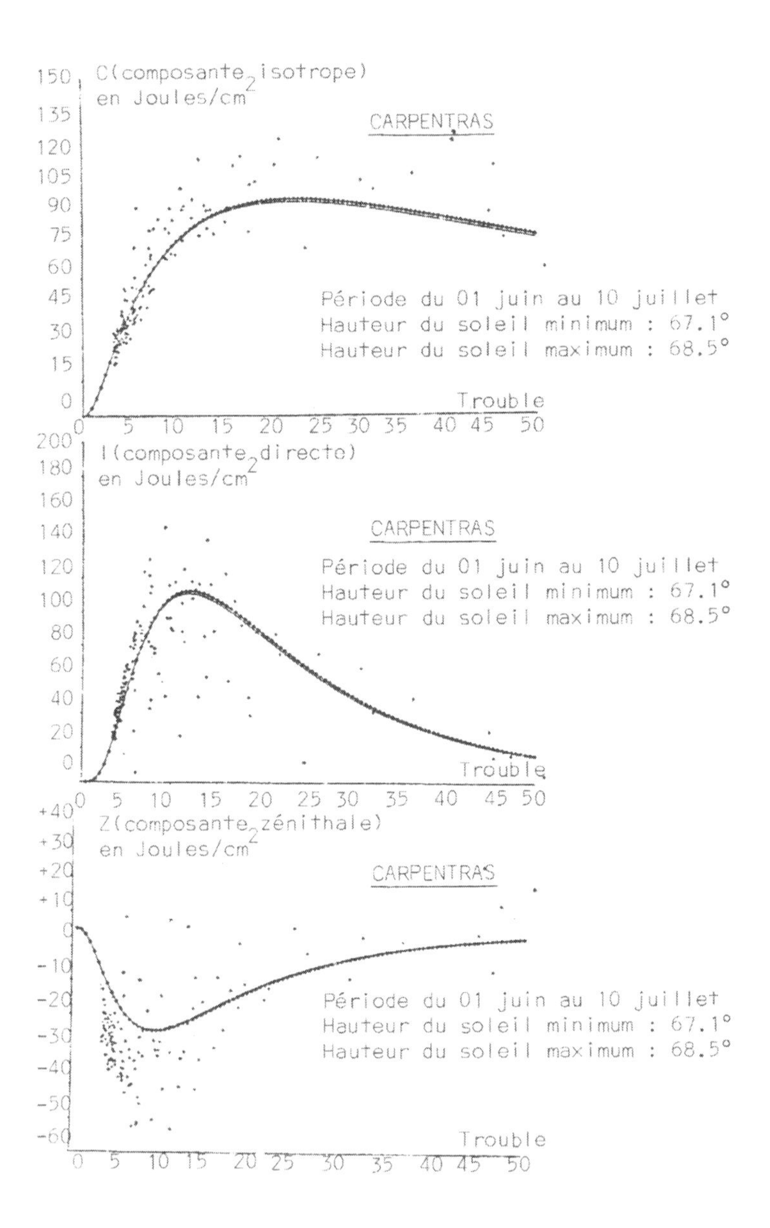

[6] Three components of the diffuse component of irradiation and their dependence on the extended turbidity factor and of sun's elevation - Data for Carpentras .
[Met. Nat]

(a)

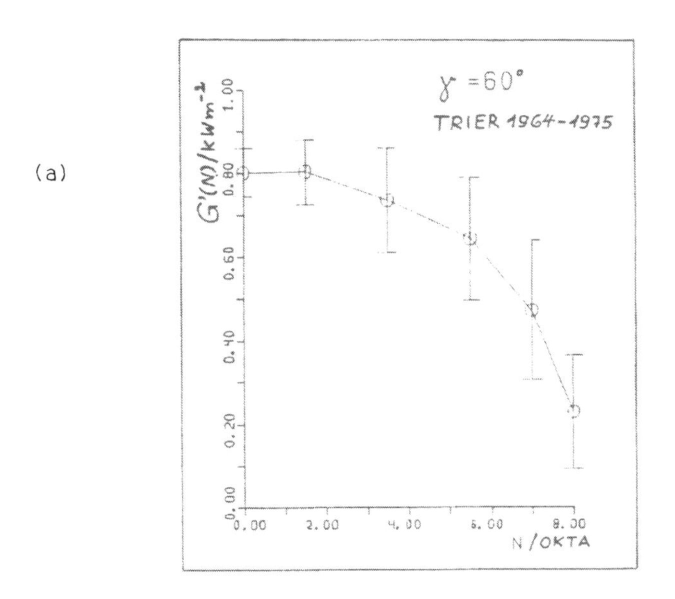

(b)

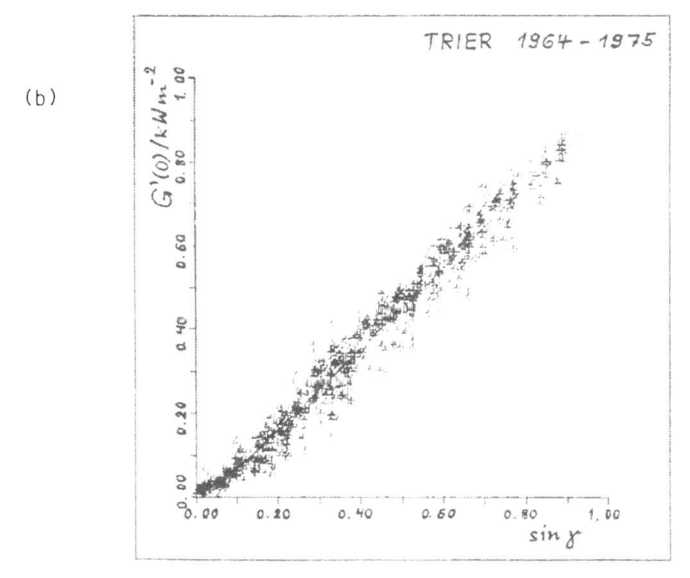

[7] Diagrams showing the variation of global irradiation $G'(N)$ as function of total cloud amount (N) with solar elevation γ as parameter (a) and the variation of $G'(0)$ as function of $\sin\gamma$ (b)
[MET.OBS, HAMBURG]

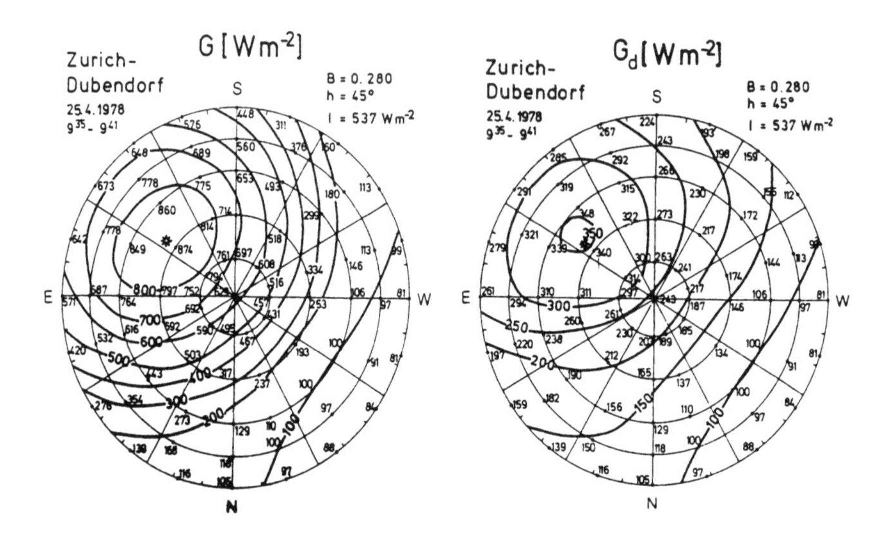

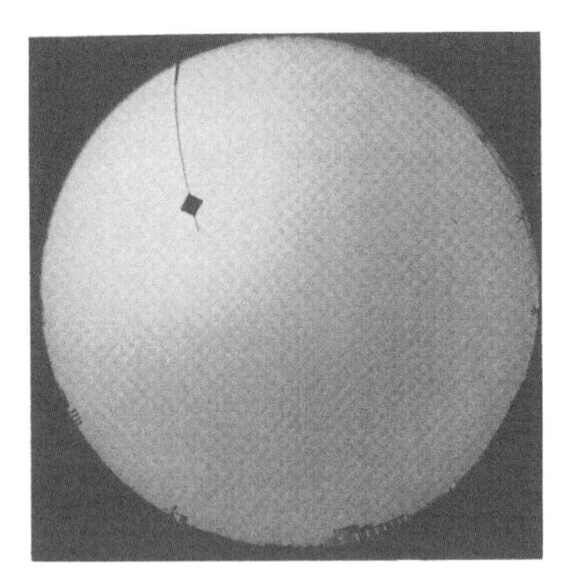

[8] Simultaneous angular distributions of the global (top left) and diffuse (top right) irradiances for differently orientated and tilted surfaces; and visualized hemisphere by the corresponding fish-eye picture (bottom) - clear sky conditions, strong turbidity, Data obtained with the portable meteorological instrument package from the Swiss Meteorological Institute, Zürich.

Action 3.3 - STATISTICAL ANALYSIS OF METEOROLOGICAL DATA

ACTION LEADER : J.A BEDEL
DIRECTION DE LA METEOROLOGIE
SERVICE METEOROLOGIQUE METROPOLITAIN
2 AVENUE RAPP
75340 PARIS CEDEX 07
FRANCE

PARTICIPANTS

- INSTITUT ROYAL METEOROLOGIQUE DE BELGIQUE
  CONTRACT ESF-003-B MR. R. DOGNIAUX

- DEUTSCHER WETTERDIENST
  CONTRACT ESF-004-D DR. F. KASTEN

- DIRECTION DE LA METEOROLOGIE
  CONTRACT ESF-005-F MR. J.A BEDEL

- THE ROYAL NETHERLANDS METEOROLOGICAL INSTITUTE
  CONTRACT ESF-006-NL DR. W.H. SLOB

- ARMINES ECOLE DES MINES DE PARIS
  CONTRACT ESF-008-F MR. B. BOURGES

- CENTRE NATIONAL DE LA RECHERCHE SCIENTIFIQUE
  CONTRACT ESF-009-F MR. J.F. TRICAUD

- UNIVERSITY OF SHEFFIELD
  PROF. J.K. PAGE

TASKS

- CUMULATIVE FREQUENCY DISTRIBUTION
- TIME SEQUENCE OF IRRADIATION
- CORRELATION BETWEEN IRRADIANCE AND OTHER METEOROLOGICAL PARAMETERS E.G. TEMPERATURE.

EXECUTIVE SUMMARY

1. Contracts

The running contracts concerning action 3.3 are grouped under the following items :

1.1 Cumulative frequency curves (CFC)

1.1.1 Computation of CFC on tilted planes with data set of hourly measurements of global radiation on tilted planes.

1.1.2 Reconstitution of CFC on tilted planes with usual network measurements (global and diffuse radiation on horizontal plane).

1.1.3 Spatial interpolation of CFC.

1.1.4 Sensivity of the CFC at the integrated time scale of the measurements.

1.2 Time sequence of irradiation

1.3 Correlation between irradiance and other meteorological parameters e.g. temperature

An intermediate report of action 3.3 is attached. Table 1 gives the general information concerning the 7 processing contracts of action 3.3.

2. Work performed and results

The main results of the work performed in 1980 and 1981 are :

2.1 Comparison of the CFC calculated with actual irradiance and with the mean hourly irradiance, for Hamburg : Dr. KASTEN's result.

2.2 Maps of the number of sequential days with a duration of sunshine

- equal to 0
- less than 0.5 hour
- less than 1 hour.

and with a statistical return period of

- once every 5 years
- three times every year.

The maps are available for France :
Results of the "Direction de la Météorologie".

2.3 Processing and presentation methods concerning the persistence of the global radiation within selected thresholds and selected time intervals, and examples for Hamburg : Dr. KASTEN's results.

2.4 Presentation methods of the correlations between irradiation and other meteorological parameters.

2.4.1 At the monthly level (monthly sums and montly means) between irradiation and temperature for Hamburg : Dr. KASTEN's results.

2.4.2 Development of a method to present the correlations at the hourly level (hourly data) by a mathematical parameterization of hourly and daily variations of global radiation and temperature : Mr. DOGNIAUX's result.

2.4.3 Mapping presentation of the correlations at the hourly level. Results of the Direction de la Météorologie.

3. Difficulties encountered

3.1 Mr. DOGNIAUX met some administrative difficulties for the approbation of his contract : no any positive information received from the Belgian Government concerning the Belgian contribution to project F.

3.2 For his work, Mr. BOURGES needs calculation models of the global radiation on tilted planes at the hourly level. The development of such models by group 3.2 of Project F, will be provided "at a later stage" (see : strategy paper - Action 3.2).

So Mr. BOURGES adopts the following working plan :

- collect of data measured on tilted planes and computation of the CFC
- computation of the CFC on tilted planes from different data (global, horizontal, diffuse..) for the same case, and comparison with the preceding.
- choice of empirical relationships for the reconstitution of diffuse irradiance on tilted planes.
- computation on long-term CFC's of solar radiation on tilted planes for few European stations.

4. Future work

4.1 the work about development of new methods of processing and presentation of solar meteorological data has been completed (see : point 2 of the present report), except :

4.1.1 Reconstitution methods of CFC on tilted planes with usual measurements (see point 3.2 of the present report).

4.1.2 Statistical presentation of the time sequences of irradiation at the hourly level. Studies are in progress in the "Direction de la Météorologie" concerning this item. Results will be available in 1982 and presented to the group 3.3.

4.2 In 1982, the main work of the contractors of action 3.3 will be the production of results with available methods (CFC and methods developed in 1980 and in 1981 : see point 2 of the present report) and using new data sets of measurements of direct and global radiation. The expected results are :

4.2.1 CFC calculated with new data sets of hourly values of direct and global radiation on tilted surfaces. These measurements are made at : Odeillo, Hamburg, Trappes, Carpentras and Cabauw. Details concerning those measurements are given in table 2.

4.2.2 Comparison of CFC's calculated with actual irradiance or 6 minute data and with the mean hourly irradiance, for Hamburg, Trappes, and Carpentras.

4.2.3 Correlation between solar irradiation, temperature and wind speed for 8 English selected stations.

4.2.4 Calculations of the mathematical parameterization of hourly variations of global radiation and temperature for 2 Belgian stations.

5. Other activities of Group 3.3

5.1 A new action leader has been appointed for this group; as Mr PERRIN DE BRICHAMBAUT was not able to go on with the leadership of the group, together with his other professional activities. The new action leader is Mr. BEDEL.

5.2 A meeting has been organized in Paris (17-19 December 1980). During this meeting 2 items have been discussed :

- Correlation structure of meteorological data especially between irradiation and temperature.
- Time structure of the solar radiation data. The needs of the users and some statistical methods of data processing have been investigated. The conclusions of the meeting have been presented in the coordination meeting of contractors (Brussels, 1-2 April 1981).

5.3 A new contract has been approved during the expert meeting of Group F (Brussels, 3 April 1981). The contractor is Professor J.K. PAGE. This new contract has been headed :
"Studies of relationships between inclined surface irradiation, temperature and wind speed for the North of Europe".
The work will begin on January 1st, 1982.

6. Strategy papers

6.1 Tasks of Action 3.3 : no remark.

6.2 Methodology

6.2.1 Cumulative frequency distributions : inclined surfaces, December 1982 :
Because of the difficulties encountered by Mr. BOURGES (see point 3.2 of the present report), it is not sure that the work will be achieved in December 1982.

6.2.2 Correlation with meteorological parameters, Dec. 1982 : no remark.

6.3 Products

6.3.1 Time sequence of irradiation, daily values, inclined planes, for 49 stations, Dec. 1982 - no remark.

6.3.2 Correlation between irradiation and temperature, June 1983.
A special report should be written on the same topic (see point 6.4.2). Nevertheless perhaps it is not necessary to write 2 reports concerning the same matter.

6.4 Special reports

6.4.1 Report on cumulative frequencies, hourly data, horizontal and inclined surfaces, December 1982.
A provisional summary of this special report is given in Appendix A.

6.4.2 Report on correlation between irradiance and other meteorological parameters, June 1983 (see point 6.3.2).

7. Conclusion

7.1 Work is getting on normally and in accordance to the prospects of the strategy of Project F, except :

7.2 Work concerning the reconstitution of the CFC on tilted planes (see point 3.2).

7.3 Concerning the strategy paper, I propose to retain only one of the 2 reports about the correlation between irradiance and other meteorological parameters (see point 6.3.2 and 6.4.2).

INTERMEDIATE REPORT (1st NOVEMBER 1981)

1. Cumulative frequency curves (CFC)

The results of the previous programme of project F have shown the usefulness of CFC for solar energy applications. The work in progress about CFC concerns the 4 following items :

- Computation of CFC on tilted planes with data set of hourly measurements of global radiation on tilted planes.

- Reconstitution of CFC on tilted planes with usual network measurements (global and diffuse radiation on horizontal planes).

- Spatial interpolation of CFC.

- Sensitivity of the CFC at the integrated time scale of the measurements.

1.1 Computation of CFC on tilted planes with data set of hourly measurements of global radiation on tilted planes

Experimental stations of solar radiation measurements on inclined surfaces have been developed in a few countries of the Community. 4 contracts intend to analyse those new data sets in terms of frequency distributions. Those new analyses complete CFC's, calculated by Mr. BOURGES in the first research programme.

This item affects 4 contracts :

- ESF 009 F   Mr. TRICAUD
- ESF 004 D   Mr. KASTEN
- ESF 006 NL Mr. SLOB
- ESF 005 F   Mr. BEDEL

The work needs no further development, the method of computation of CFC is the method developed by Mr. BOURGES in the first research programme. Details concerning the data sets are given in table 2.

The work will be performed later, with a duration of measurements as long as possible. The results are not available for the present time. Only Mr. KASTEN has presented intermediate results.

The difficulties were related to data acquisition systems; they have been overcome now.

1.2 Reconstitution of CFC on tilted planes with usual measurements

The aim is to develop methods of reconstitution of CFC on tilted planes. This item concerns the contract :

ESF 008 F Mr. BOURGES.

At present, there is no existing model approved by the CEE for the calculation of the global radiation on tilted surfaces at the hourly level.

So Mr. BOURGES adopted a working plan (see point 3.2 of the executive summary) using empirical methods for the reconstitution of the global irradiation on tilted planes.

1.3 Spatial interpolation of the CFC

Mr. BOURGES has proposed a method of spatial interpolation of the CFC on horizontal plane. The extent of this method to the CFC on tilted planes would be very useful, but the necessary data are too scarce at present.

We have no proposal concerning this point, only Mr. SLOB will make a contribution concerning this point. He proposes to study the relations between the CFC, the sunshine duration and the turbidity in Cabauw, but it will not be possible to extend his results and to propose a method of spatial interpolation.

1.4 Sensitivity of the CFC at the integrated time scale of the measurements

The aim is to estimate the differences between the CFC calculated with short interval integrated data.

This item concerns 2 contracts :

- ESF 004 D Mr. KASTEN
- ESF 005 F Mr. BEDEL.

The study needs no further development.

Mr KASTEN has compared the CFC calculated with actual irradiance and with the mean hourly irradiance, using Hamburg's data (see table 2). The comparison is made only for durations during which the radiation is over 600 w/m^2, 400 w/m^2 and 100 w/m^2.

Results have been presented at the coordination meeting of contractors (Brussels, 1-2 April 1981).

In France, the comparison will concern the CFC calculated with hourly data and with 6 minute data. The work will be performed in 1982, using data from Carpentras and Trappes (see table 2).

1.5 The "Ecole des Mines de Paris" has compared various methods of estimating the useful energy of solar collectors :

- utilizability computed from CFC
- utilizability from Klein
- utilizability from Collares-Pereira
- utilizability computed from empirical formulas

The results of this comparison are available (work performed in 1981). This work will be integrated into the action 3.3.

2. Time sequence of irradiation

Statistical analysis of time sequences of irradiation are very useful, especially for the computation of the storage system.

This point concerns 3 contracts

- ESF 004 80 D Mr. KASTEN
- ESF 005 81 F Mr. BEDEL
- ESF 009 81 F Mr. TRICAUD

2.1 Mr. KASTEN has studied the persistance of the radiation within selected thresholds (600, 400, 300 and 100 w/m^2) and within selected time intervals (5,15,30, 60, 120 and 240 minutes). His method has been exposed at the coordination meeting of contractors in Brussels (1-2 April 1981).

2.2 An analysis of the sequences of consecutive days with a duration of sunshine

- equal to 0
- less than 0.5 hour
- less than 1 hour

has been performed for 50 stations in the "Direction de la Météorologie". Maps of the number of consecutive days with bad conditions of sunshine duration and with a statistical return period of

- once every 5 years
- 3 times every year

are available.

A statistical study of the time sequences of irradiation at the hourly level is in progress in "Direction de la Météorologie".

2.3 Mr. TRICAUD will use the statistical methods developed by the "Direction de la Météorologie". He will work the data of Odeillo (see table 2).

3. Correlation between irradiance and other meteorological parameters

The correlation between solar radiation and temperature or other meteorological parameters (wind speed, humidity...) are very important for some solar energy applications, especially for those concerning the passive systems. The problem is complex and should be studied for different time scales (month, day, hour). This point concerns 4 contracts :

- ESF 006 NL Mr. SLOB
- ESF 003 B Mr. DOGNIAUX
- ESF 005 F Mr. BEDEL
- ESF UK Mr. PAGE.

The following methods and results are available :

3.1 Mr. KASTEN has shown a correlation between the montly means of temperature and the monthly sums of solar radiation in summer, for Hamburg, and, an anti-correlation in winter.

3.2 Mr. DOGNIAUX has developed a method to present the correlations between meteorological parameters by mathematical functions giving the hourly (and daily) variations of the meteorological parameters. Those functions can be calculated for selected days representative of some characteristic weather conditions (overcast days - cold days ...).

3.3 Mr. BEDEL proposes mapping presentations of the correlations between solar radiation and temperature, at the hourly level. Those presentations have been exposed at the coordination meeting of contractors in Brussels (1-2 April 1981).

TABLE 1

CONTRACTS	ESF-003 IRMD DOGNIAUX 30.6.83	ESF-004 DWD KASTEN 31.12.82	ESF-005 DMN BEDEL 30.6.83	ESF-006 RNMI SLOB	ESF-008 ARMINES BOURGES 30.6.83	ESF-009 CNRS TRICAUD 30.6.83	SHEFFIELD PAGE 30.6.83
1. CFC							
1.1 COMPUTATION WITH HOURLY DATA SET		X	X	X		X	
1.2 RECONSTITUTION OF CFC ON TILTED PLANES WITH USUAL DATA					X		
1.3 SPATIAL INTERPOLATION							
1.4 SENSITIVITY AT THE INTEGRATED TIME SCALE OF MEASUREMENTS		X	X				
2. TIME SEQUENCES OF IRRADIATION		X	X			X	
3. CORRELATION BETWEEN IRRADIATION AND OTHER PARAMETERS	X		X	X			X

TABLE 2

NEW DATA SETS
OF MEASUREMENTS OF GLOBAL RADIATION
ON TILTED PLANES IN E.E.C.

CONTRACT	STATION	BEGINNING OF THE MEASUREMENTS	MEASUREMENTS	
ESF-009 F	ODEILLO	1979	DIRECT AND GLOBAL ON FOLLOWING PLANES -90° SOUTH,EAST AND WEST -42°5 SOUTH -HORIZONTAL	6 minute DATA
ESF-004 D	HAMBURG	1980	GLOBAL ON FOLLOWING PLANES -90° SOUTH -54°6 SOUTH -30° SOUTH -HORIZONTAL	ACTUAL IRRADIANCE
ESF-006 NL	CABAUW	1979	DIRECT AND GLOBAL ON FOLLOWING PLANES -90° EAST, SOUTH,WEST, NORTH -67°5 SOUTH -45° EAST, SOUTH-EAST, SOUTH, SOUTH-WEST, WEST -22°5 SOUTH -HORIZONTAL	6 minute DATA
ESF-005 F	CARPENTRAS AND TRAPPES	1979	DIRECT AND GLOBAL ON FOLLOWING PLANES -90° EAST, SOUTH, WEST, AND NORTH -45° SOUTH -HORIZONTAL	6 minute DATA

APPENDIX A

PROVISIONAL SUMMARY OF THE SPECIAL REPORT ON CUMULATIVE FREQUENCIES

1. GENERAL PRESENTATION OF THE CF AND CFC.
2. EXAMPLE OF APPLICATIONS OF THE CFC TO ESTIMATE THE USEFUL ENERGY FOR SOLAR COLLECTORS.
3. METHOD OF COMPUTATION OF THE CFC WITH HOURLY DATA OF GLOBAL RADIATION.
4. CFC ON HORIZONTAL PLANES FOR 29 EUROPEAN STATIONS.
5. METHOD FOR ESTIMATING THE CFC ON HORIZONTAL PLAN WITH THE MONTHLY MEANS OF SUNSHINE DURATION.
6. CFC ON TILTED SURFACES (CABAUW, TRAPPES, CARPENTRAS, ODEILLO,...).
7. METHOD FOR THE RECONSTITUTION OF THE CFC ON TILTED PLANES.
8. CFC ON TILTED PLANES FOR A SELECTION OF EUROPEAN STATIONS.

APPENDIX B

THE MAIN RESULTS AVAILABLE :

- COMPARISON OF THE CFC CALCULATED WITH ACTUAL IRRADIANCE AND WITH THE MEAN HOURLY IRRADIANCE, FOR HAMBURG. (fig. 1)

- PERSISTENCE OF THE GLOBAL RADIATION WITHIN SELECTED THRESHOLDS AND SELECTED TIME INTERVALS, FOR HAMBURG. (fig. 2)

- CORRELATION (AND ANTICORRELATION) BETWEEN THE MONTHLY SUMS OF IRRADIATION AND THE MONTHLY MEANS OF TEMPERATURE. (fig. 3a and 3b)

- DENSITY OF PROBABILITY : TEMPERATURE AND GLOBAL RADIATION AT THE HOURLY LEVEL . (fig. 4)

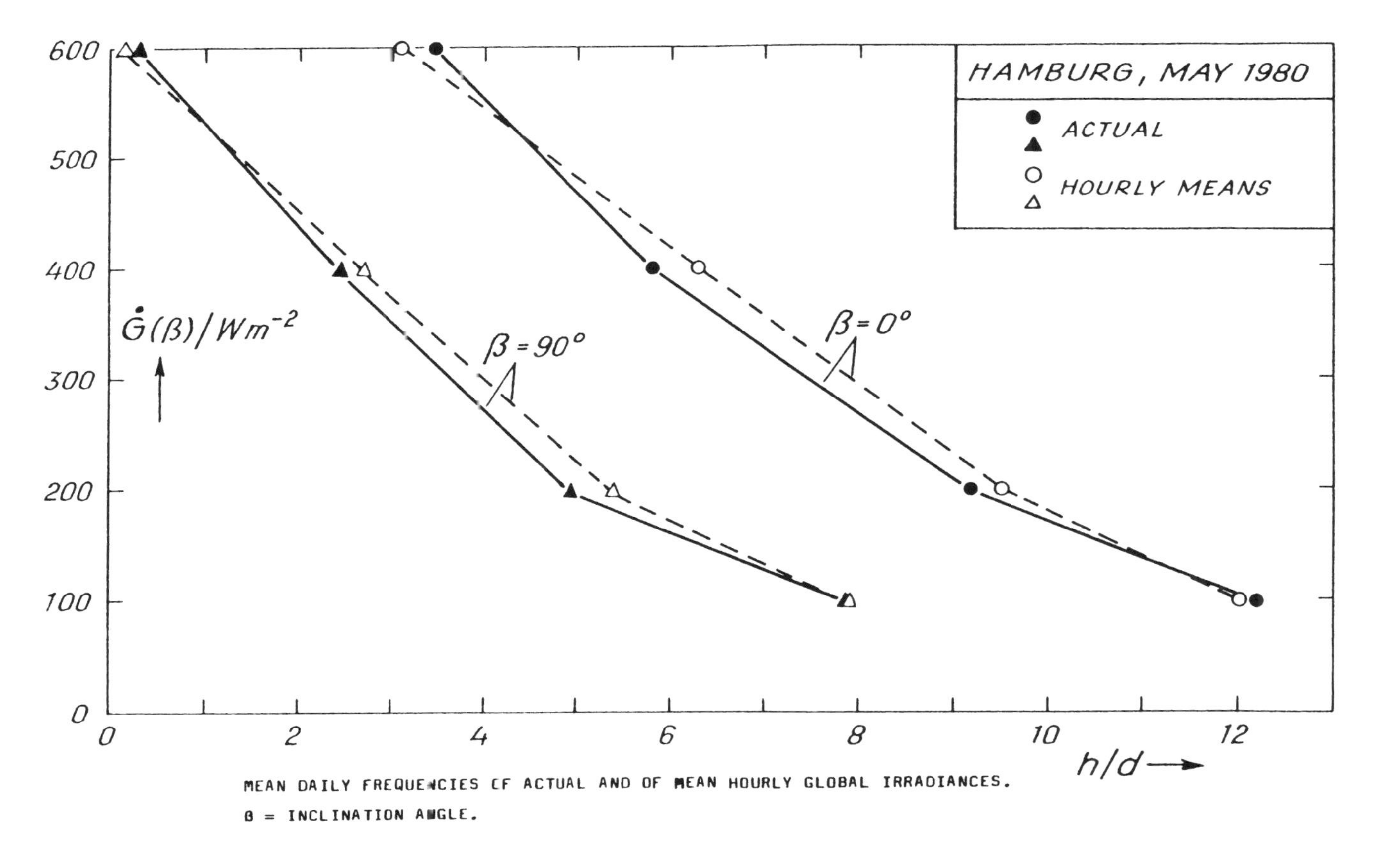

Fig. 1 - Comparison of the CFC calculated with actual irradiance and with the mean hourly irradiance, for Hamburg.

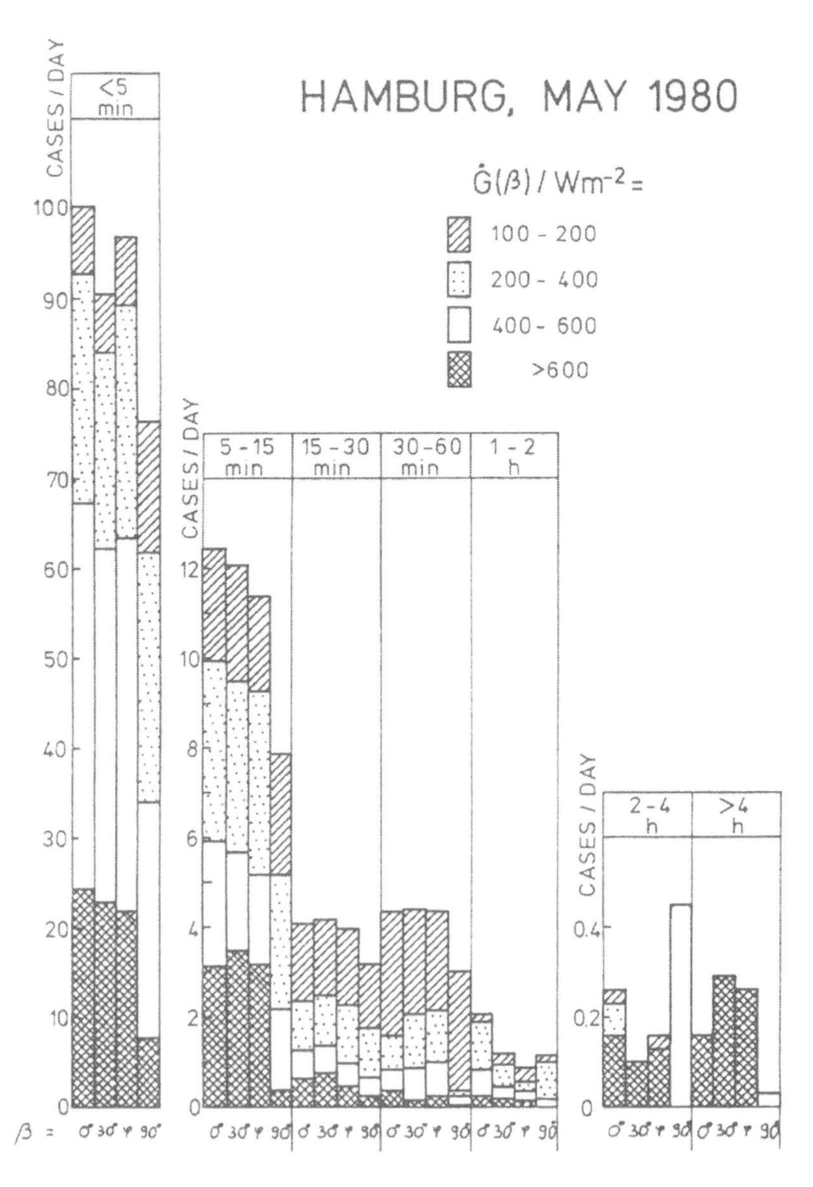

MONTHLY MEANS OF CASES PER DAY WHEN THE GLOBAL IRRADIANCES ON INCLINED PLANES, $\dot{G}(\beta)$, PERSIST WITHIN GIVEN RANGES FOR GIVEN TIME INTERVALS.
β = INCLINATION ANGLE.

Fig. 2 - Persistence of the global radiation within selected thresholds and selected time intervals, for Hamburg

(a)

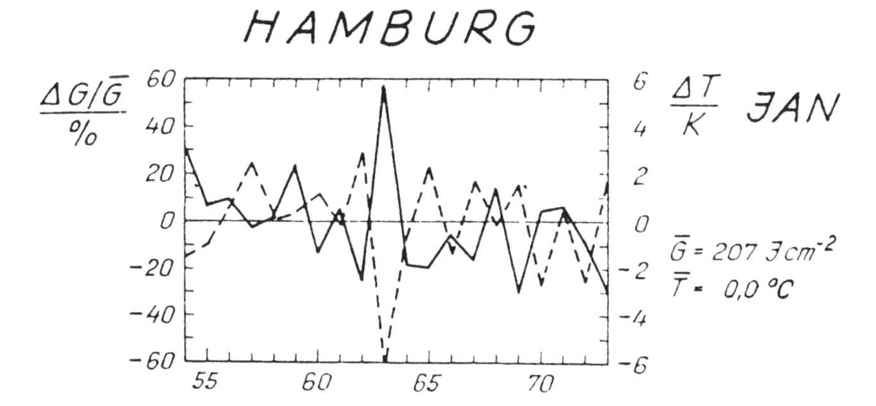

(b)

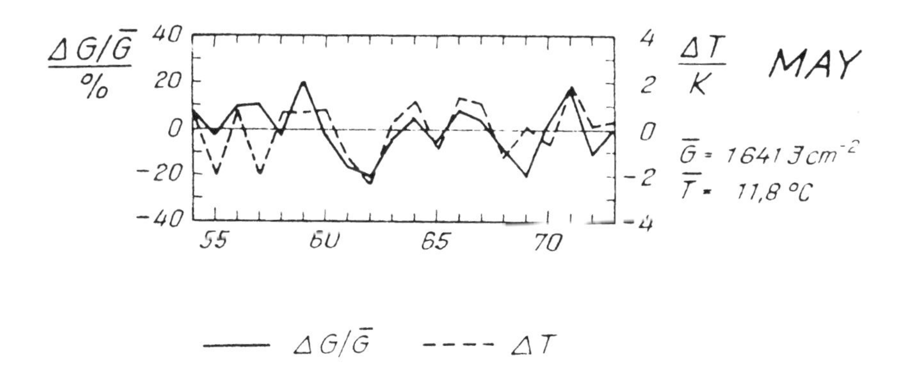

YEAR TO YEAR VARIATION IN EACH MONTH OF DAILY GLOBAL RADIATION AND TEMPERATURE. DEVIATIONS OF INDIVIDUAL MONTHLY MEANS FROM THEIR 20 YEAR MEANS.

Fig. 3 - Correlation (and anticorrelation) between the monthly sums of irradiation and the monthly means of temperature

Density of probability (of the hourly data)

View of the fonction

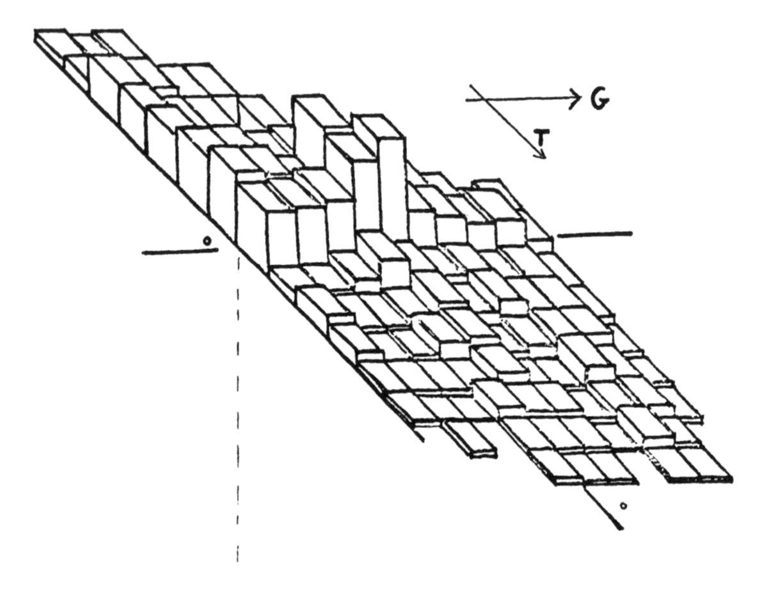

Fig. 4 - Density of probability : temperature and global radiation at the hourly level

APPENDIX C

DEVELOPEMENT OF A METHOD OF PARAMETERIZATION OF THE DAILY VARIATIONS OF THE TEMPERATURE (T) AND THE GLOBAL RADIATION (G).

(SEE M. DOGNIAUX'S REPORT - 1/2 APRIL 1981).

$$T = A + B \cos wt + C \cos 2\,wt + D \sin wt + E \sin 2wt$$

t : BEING EXPRESSED IN HOUR
w : $2\,\pi/24$

A, B, C, D, E, ARE GIVEN IN FUNCTION OF THE CLASSES OF DAYS.

$$G = A.\exp\left(- \frac{B}{X_o^2 - X^2} - \frac{C}{X_o^4 - X^4} \right)$$

G IS EXPRESSED IN $W.m^{-2}$

Xo THE HALF LENGTH OF THE DAY EXPRESSED BY THE HOUR ANGLE OF THE SUN AT SUNRISE OR SUNSET.

X THE TIME.

A, B, C, ARE GIVEN IN FUNCTION OF THE CLASSES OF DAYS.

Action 3.4 - SENSITIVITY ANALYSIS OF THE USEFUL ENERGY OUTPUT FOR SOLAR CONVERTERS WITH REPECT TO QUALITY AND COMPLETENESS OF METEOROLOGICAL DATA SETS

Action leader: Until September 81: Prof. Pilatte, Faculté Polytechnique de Mons.
Since October 81 without action leader.

Author of this paper: H.R. Koch, KFA Jülich, Institut für Kernphysik

Participants in the action:
- Faculté Polytechnique de Mons, without contract
  (Prof. Pilatte)
- Ecole des Mines de Paris
  (contract June 80 to Dec. 81)
  (Mr. B. Bourges)
- KFA Jülich (contract no. ESF-022-80D(B), Sept. 80 to Dec. 81)
  (H.R. Koch)

Summary

The subject of this study is to determine, how precisely the energy, produced by solar collectors, can be calculated, when reduced meteorological data sets are used.

The effect of time steps from 1 minute to one hour was studied, for the energy of a flat plate collector and for an evacuated tubes collector, for different inlet temperatures. Results were obtained for 2 data sets of six days, characterized by fluctuating and by permanent sunshine respectively. The applicability of a collector efficiency, which depends only on the reduced temperature T^*, for the calculation of the all day performance of a flat plate collector, was proved experimentally. This justifies the use of radiation energy distributions ("RED"-functions) for collector heat output calculations. These functions were calculated for the Belgian Reference Year and energies of the CEC4 collector were calculated. The results are in "good" or "sufficient" agreement with energies obtained with the help of detailed simulation calculations. The applicability of the cumulative frequency curves was studied with a 6-days data set: the heat output of the CEC4 collector was calculated for different mean air temperatures. The mean values for the periods 9-15 h lead to good agreement with simulation calculations.

Meetings and reports concerning action 3.4

Participants in the meetings:

- Faculté Polytechnique de Mons (Prof. Pilatte)
- Ecole des Mines de Paris (B. Bourges, Mr. Watremez)
- KFA Jülich (H.R. Koch, W. Grüter (Jan. 80), W. Scheller (Nov. 81)

8 Jan 1980, Jülich	Definition of the tasks of the action
31 Jan 1980	Report "Description of the tasks" by Prof. Pilatte
17 March 1980, Mons	Coordination meeting action 3.4
5,6 May 1980, Jülich	CEC-F coordination meeting, reports
17 to 19 Dec 1980, Paris	Workshop in Paris on "Time and correlation structures of meteorological and radiometric data for solar energy applications". Reports given by the three participants of action 3.4
5 March 1981, Mons	Coordination meeting action 3.4
1,2 April 1981, Brussels	Contractors meeting, reports
10 Nov 1981, Paris	Coordination meeting action 3.4
20 Nov 1981, Brussels	Review meeting, report
Jan 1982, Mons	Coordination meeting action 3.4 foreseen

1. Introduction

Systems, which make use of solar energy, can be most perfectly designed using complete short interval meteorological data sets, in which the correlation between the various parameters and their time sequence are preserved. The use of reduced meteorological data sets makes the computations easier, but the results may be less valuable. The subject of the work presented in this paper is a study on the precision of the heat output calculations of solar converters, when different types of reduced weather data are used. The collector heat output is the energy, which is fed into the pipes leading in general to a storage tank. Of course this energy does not determine alone the quality of a solar system. But the solar converter is, among the components of a solar system, the one, whose performance is most strongly related with weather conditions. The fraction of the collector energy, which is finally used by the energy consumer and which replaces other energy resources, this fraction can only be computed, regarding the interaction of all the components of the system and must be the subject of further research. In the present study we have computed collector heat outputs, assuming in a 1st step the simplest action of the system on the collector: we regard constant inlet temperatures and constant flow rates, leading to operating temperatures of the collector, which vary only by some degrees.

2. How to judge the validity of reduction methods

The reduced data sets are used to calculate the heat output of a collector. These results are compared with the real heat output. In order to classify the differences which may appear, we must have a quality scale. We have learned from the round robin collector test of the IEA (1) and the EC (2) that the variance of η-determinations is in the order of 5 % for test procedures with defined meteorological conditions. Considering these differences one may define, of course somewhat arbitrarily, the following quality scale for the computation of the heat output of collectors of long periods, during which all types of weather conditions may appear:

deviation	quality
<10 %	good
<20 %	sufficient
>20 %	questionable

The reference heat output of a collector, which is used as a standard in the comparison of computed results, can be obtained through precise measurements or through a detailed simulation calculation, based on the short interval weather data and the physical properties of the collector. We may assume that the result of a detailed simulation calculation is by far more precise than results of simplified methods, provided that the correct collector parameters are known. The simulation programmes used for the work in the action 3.4 are the COL2 programme of the Faculté Polytechnique de Mons (FPM) and the Minersol programme of the Ecole des Mines (ECM), and a simulation programme developed at Jülich. The COL2 and the Jülich programme were applied to solve identical problems. The results coincided satisfactorily. However in the following only the COL2 results are used, as this programme is the more fundamental one.

3. Data reduction methods

3.1 General remarks on the applied methods

The methods of data reduction, applied within action 3.4 are shown schematically in fig. 1. The transformation of datasets D0 to D1 and D2 is a simple integration, leading to longer time steps. The transformations D0, D1, or D2 → CFC or RED are more complex reductions, which were described previously (3,4). The Cumulative Frequency Curves ("CFC") are well established within the programme F and need no explanation at this place. The "RED"-function (Radiation Energy Distribution) is a simplified version of the MURD-function. The latter is applied in Switzerland by Kesselring and his coworkers (5). The method is based on the assumption that the collector efficiency η is in good approximation only a function of the reduced temperature $T^* = U_0(T_c - T_a)/G$ ($U_0 = 10$ W/(m2·K), T_c = mean collector temperature, T_a = temperature of ambient air, G = irradiance on the collector plane). The value of the RED-function for a certain $T^{*\prime}$ is the sum of G for all intervals, for which T^* falls within a slice between $T^{*\prime}$ and $T^{*\prime}+\Delta T^*$. For normalization this sum is divided by ΔT^*. The collector heat output for the period, which was used for the computation of the RED-function, is given as $Q(T_c) = \Sigma\, RED(T^*) \times \eta(T^*)\cdot\Delta T^*$, where the summation comprises all T^*-slices. This mode of proceeding is applicable for fixed collector temperatures T_c. Figure 2 illustrates, how Q is derived from the RED- and η-curves. It can be shown easily that the same heat output is obtained when the most primitive simulation calculation is performed, using the original data and the efficiency $\eta(T^*)$ for each of the intervals of the investigated period:

$$Q = \sum G\cdot\eta(T^*)$$

Therefore the applicability of the RED-function can be studied through an investigation of the applicability of an efficiency η, which depends only on T^*, for all day heat output calculations. In the following limitations of the $\eta(T^*)$-concept are briefly discussed, in order to show that a careful study is needed.

Deviations between $\eta(T^*)$ and the real efficiency are introduced through the following effects:

a) Wind speed.
b) The efficiency η is only valid for the global radiation G fixed in the test procedure. The heat loss coefficient usually is given as $\alpha = a_1' + a_2'\cdot\Delta T$. The coefficient of the quadratic term of the η-curve ($\eta = a_0 + a_1 T^* + a_2\cdot T^{*2}$), derived from α, depends on G:

$$a_1 = a_1'/U_0;\; a_2 = a_2'\cdot G/U_0^2$$

This effect vanishes only if a constant heat loss coefficient is used, corresponding to a linear η-curve.
c) Radiation incident at angles $<45^0$ is overestimated.
d) The dynamical behaviour of the collector, due to its heat capacity, leads to an overestimation of the heat output.

3.2 Effects of time interval lengths

The effect of time interval lengths of meteo-data on the computed collector heat output was studied by ECM, using data sets of

- Paris (6 days, 1 minute time step, rapidly varying insolation predominates, fig. 3) and of

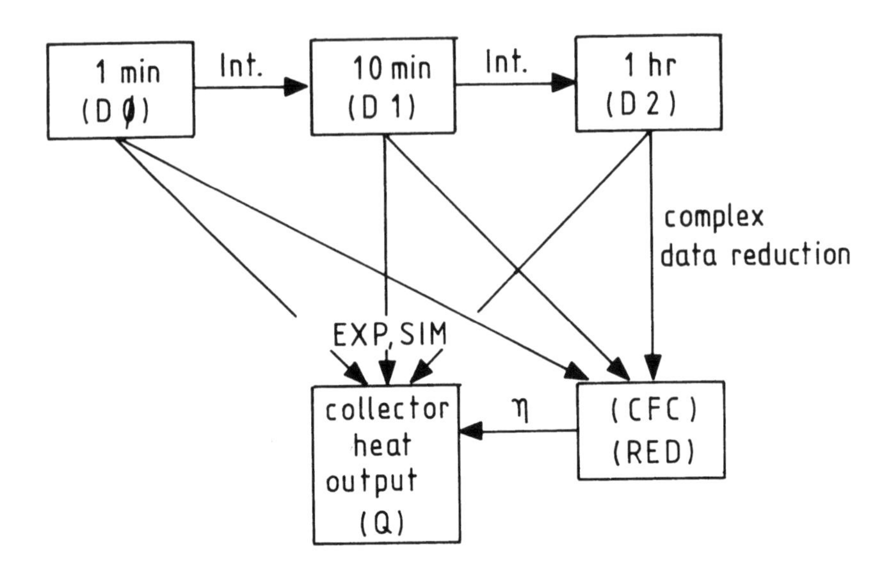

Fig. 1: Flow diagram for the applied data reduction methods

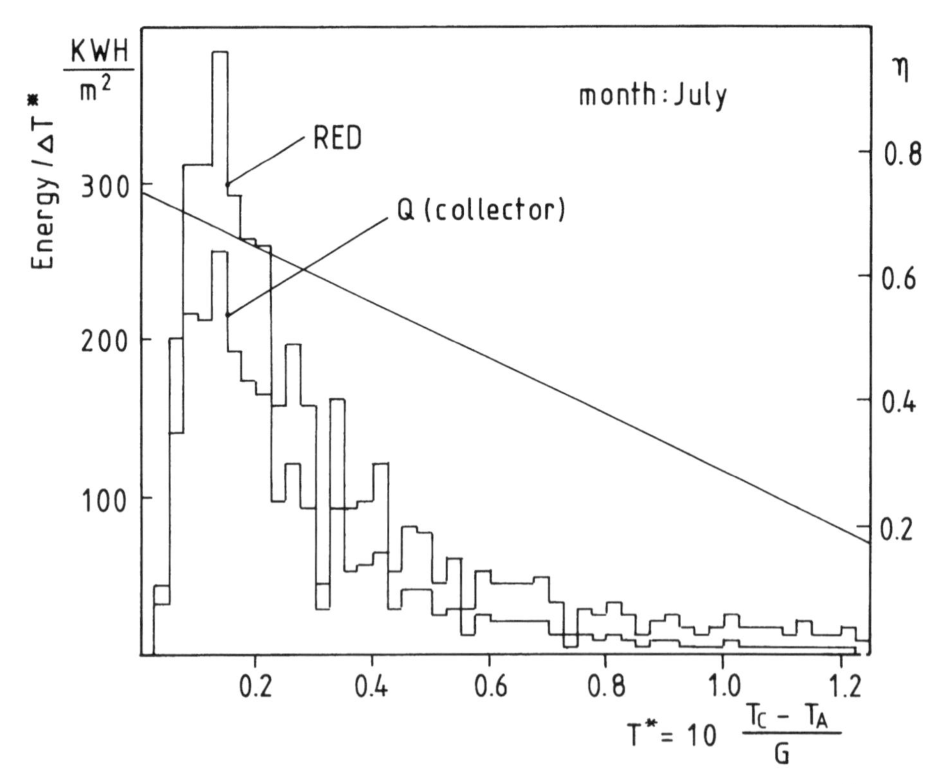

Fig. 2: Example of the application of the RED-function ($T_C = 30^0$, $Q = \Sigma\ RED \cdot \eta \cdot \Delta T^*$)

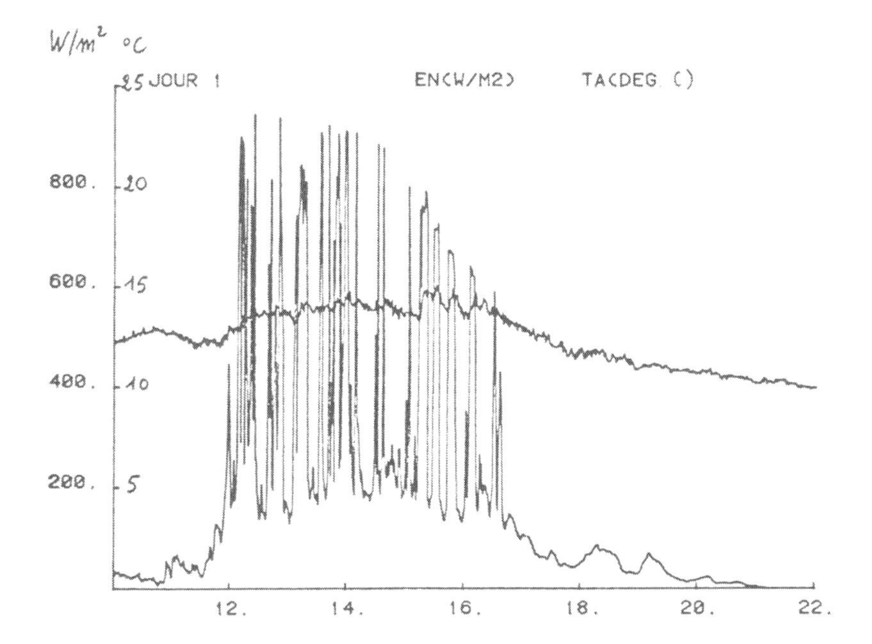

Fig. 3: Global radiation of a typical day of the Paris data

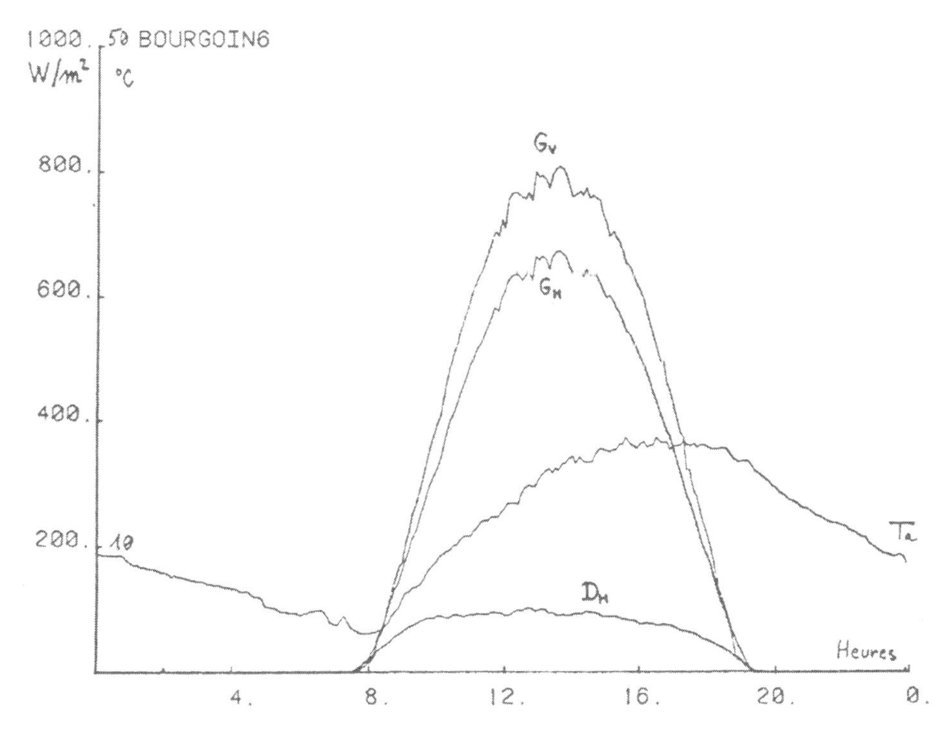

Fig. 4: Radiation (vertical, horizontal, diffuse) for a typical day of the data of Bourgoin-Jallieu

- Bourgoin Jallieu (dép. Isère, 6 days, 6 minutes time step, bright days predominate, fig. 4).

The collectors, which were studied are the CEC4 collector (Drucollector, flat plate) and an evacuated tubes collector (Mazda). In the figures 5a,b and 6a,b the ratio ε of computed collector energies to a computed reference value is shown as a function of the time interval length. Referring to fig. 1, results are obtained for the lines

$$D0 \xrightarrow{SIM} Q; \ D1 \xrightarrow{SIM} Q; \ D2 \xrightarrow{SIM} Q .$$

The simulations SIM were performed with Minersol and with the simple application of the Hottel-Whillier-Bliss-formula ("HWB"):

$$Q = \int (a \cdot G + \alpha (T_a - T_c)) dt .$$

The latter relation should give results close to those obtained with $\eta(T^*)$.

The results of this investigation may be summarized as follows:

For detailed simulation calculations an increase of the time steps decreases the collector energy. For brigth days this effect is negligible, for days of rapidly alternating insolation the effect may be in the order of 10 %. For primitive simulation, using the HWB-relation, the longer time steps (1 hr) give better results than the shorter ones. For time steps of 1 hr the deviations are in all cases <10 %. These results are in complete agreement with a study which was made by the KFA within the first part of action 3.3.

3.3 Complex data reduction

3.3.1 Test of the applicability of the RED-function

As discussed in chapter 3.1, the value of the RED-function concept can be deduced from a determination of the all day validity of the efficiency $\eta(T^*)$. This question has been studied by the KFA in two ways:

a) Comparison with experiment

For a period of 9 days in August 81 the heat output of a flat plate collector (type: Viessmann Acredal-s, 2 covers) was measured at Jülich in time steps of 10 minutes and compared with $Q = \eta(T^*) \cdot G$. In the terms of fig. 1 we have compared

$$D0 \xrightarrow{EXP} Q \text{ with } D1 \rightarrow RED \xrightarrow{\eta(T^*)} Q .$$

The latter process is equivalent with

$$D1 \xrightarrow{\eta(T^*)} Q$$

for constant collector temperatures. Figures 7 and 8 show the results for the 2nd (varying insolation) and the 6th of August (permanent insolation). We observe in both cases the expected delay of the collector heat output, when the insolation changes. The deviation of 5 % in figure 8, which is most clearly seen during the long period of constant irradiance at noon, is probably due to a difference between the real $\eta(T^*)$-curve of our collector and that we used in our calculation, which had been determined for this collector type in a standard BSE collector

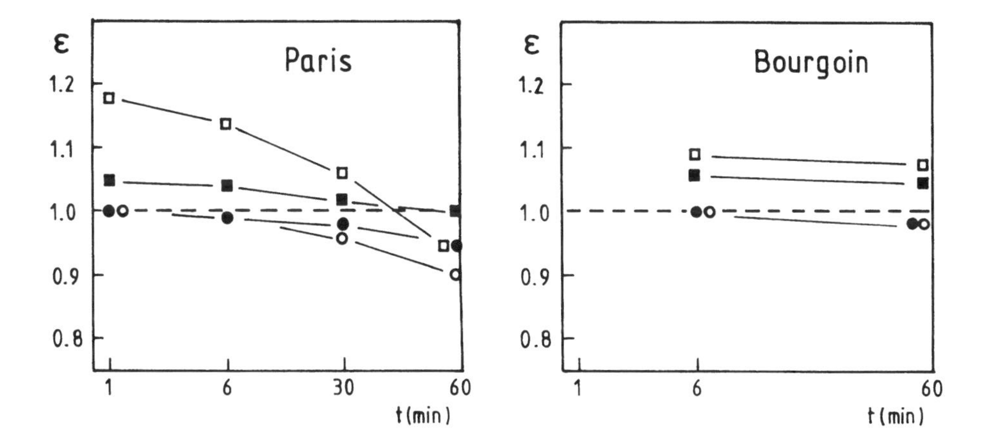

Fig. 5: Variation of computed energies with the lengths of the time steps of the meteorological data for the Dru-collector

□ HWB-Model, T_{in} = 50 °C

■ HWB-Model, T_{in} = 30 °C

○ Simulation (Minersol), T_{in} = 50 °C

● Simulation (Minersol), T_{in} = 30 °C

ε Computed energy/reference energy (t = 1 or 6 min)

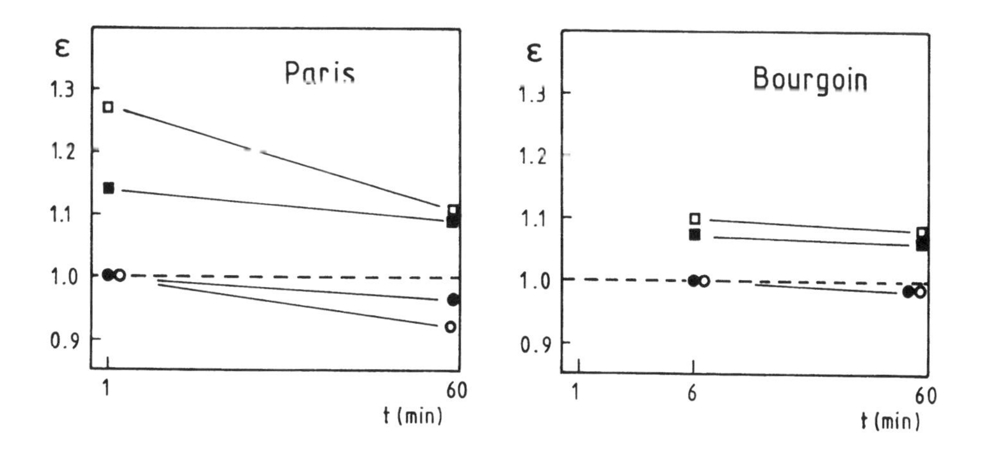

Fig. 6: Same as fig. 5, however T_{in} = 150 °C and T_{in} = 100 °C for an evacuated tubes collector

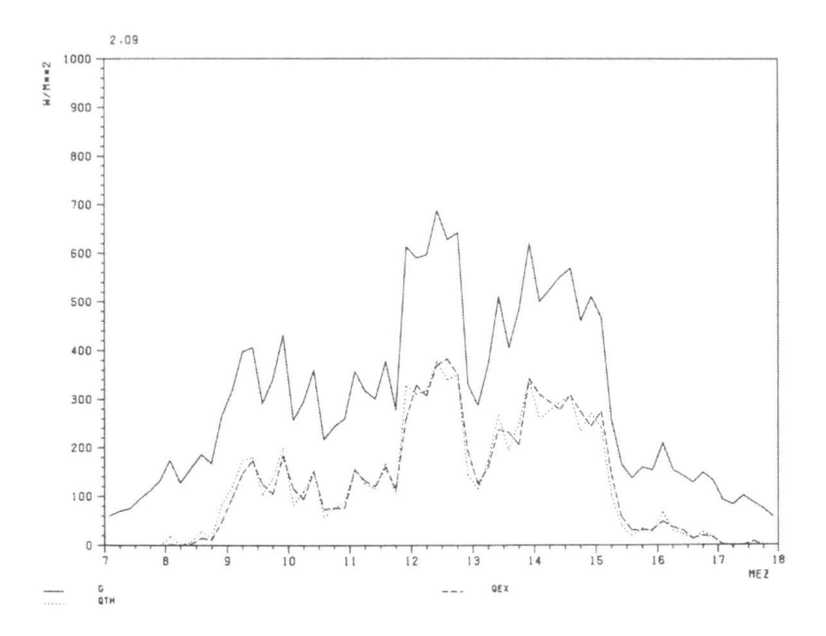

Fig. 7: Global radiation on the collector plane (——), measured collector energy (---) and computed energy $Q = \eta(T^*) \cdot G$ for August 2, 1981, at Jülich

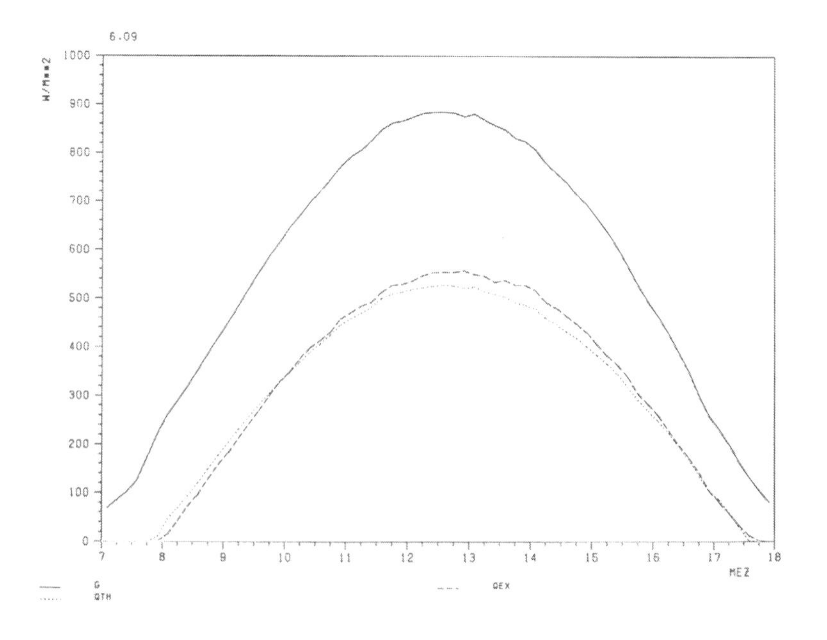

Fig. 8: Same as fig. 7 for **August** 6, 1981

test at our institute. We could have adapted of course easily the η-curve according to our measurements. But keeping to our quality scale we can very well tolerate the deviation. Table I shows the daily sums Q. The coincidence probably is due to unknown very favourable circumstances. We may not conclude that a quality scale of narrower limits can be used.

Date	Q_{exp}	Q_{com}	Q_{com}/Q_{exp}
Aug 1	0.15	0.16	1.07
" 2	1.30	1.30	1.00
" 3	2.32	2.35	1.01
" 4	2.99	2.96	0.99
" 5	3.35	3.29	0.98
" 6	3.36	3.27	0.97
" 7	0.94	0.88	0.94
" 8	0.79	0.81	1.03
" 9	1.47	1.40	0.95
total	16.67	16.42	0.99

Table I. Measured (Q_{exp}) and computed daily collector energies (Q_{com}) in KWH/M2 for a two-cover collector; $Q_{com} = \Sigma(0.68\text{-}0.34\ T^*)\cdot G$; $T_{in} = 40\ ^{o}C$

b) Comparison with detailed simulation calculations

The energy output of the CEC4-collector (Dru-collector) was determined by the KFA for the inlet-temperatures 10, 30, and 50 ^{o}C, using as meteorological input parameters the Belgian Reference Year, as defined in action 2.

The process of data reduction can be described in terms of fig. 1 by

$$D2 \rightarrow RED \xrightarrow{\eta(T^*)} Q\ .$$

As an example the RED-functions for the months of March and July are shown in figs. 9 and 10. The reference energies were obtained by Prof. Pilatte from FPM with the help of the COL2 programme (process D2 $\xrightarrow{SIM}$ Q). The monthly collector energies, obtained with the two methods are given in table II and illustrated in fig. 11. The differences amount up to 15 %. A critical comment on the significance of such differences is given in the conclusions.

The intention of the participants was that the heat output of the Dru-collector should also be calculated with the cumulated frequency method. The results will be available in December 81, as the ECM had unexpected difficulties reading the data of the Belgian Reference Year.

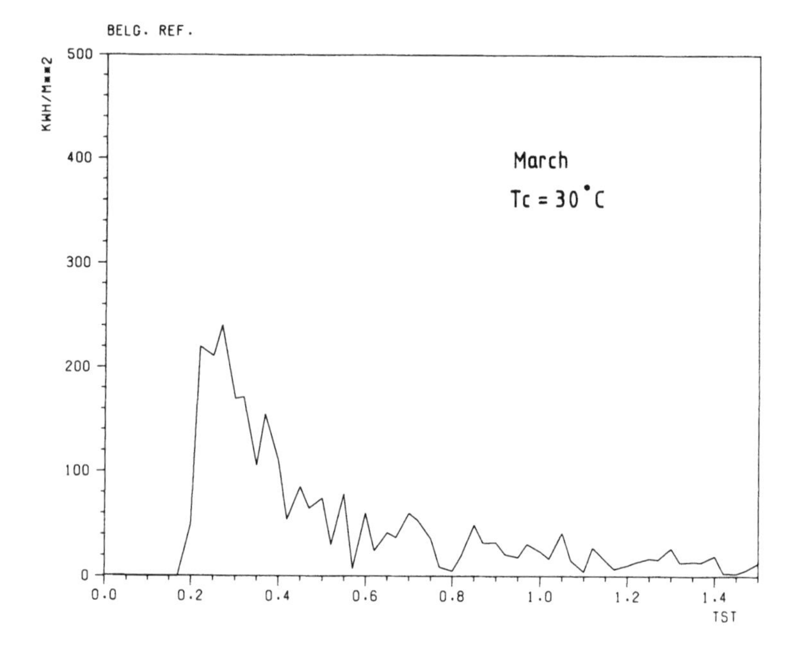

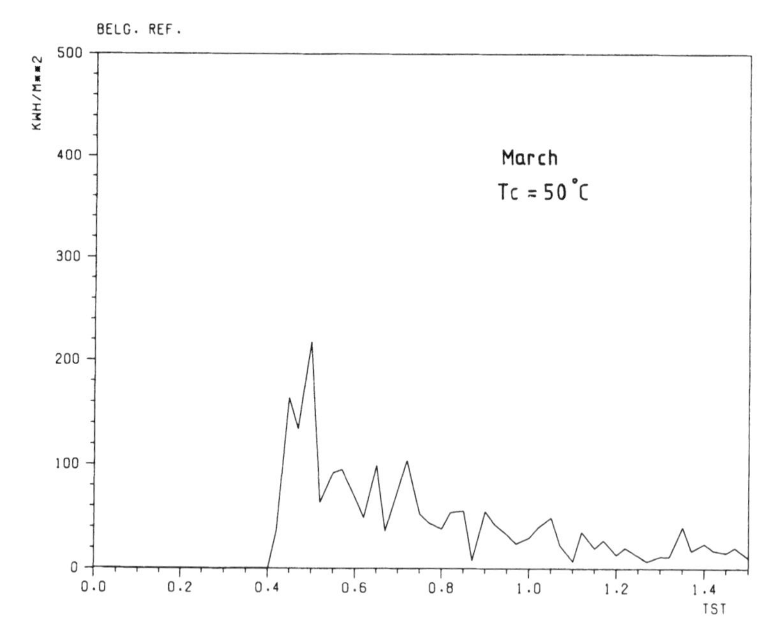

Fig. 9: Examples of RED-functions for March of the Belgian reference year for the collector temperature T_c = 30 ^{0}C (upper curve) and T_c = 50 ^{0}C (lower curve). Collector slope: 51 ^{0}C south.

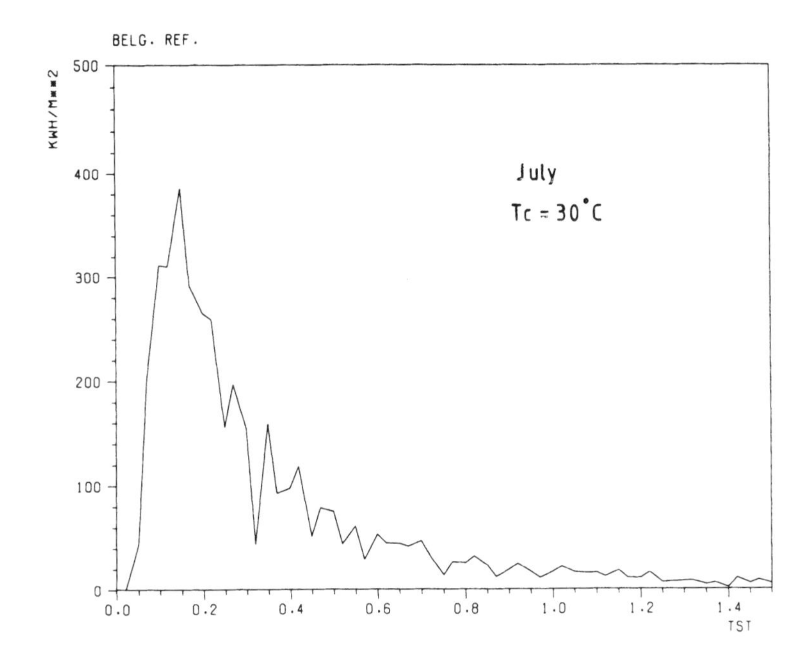

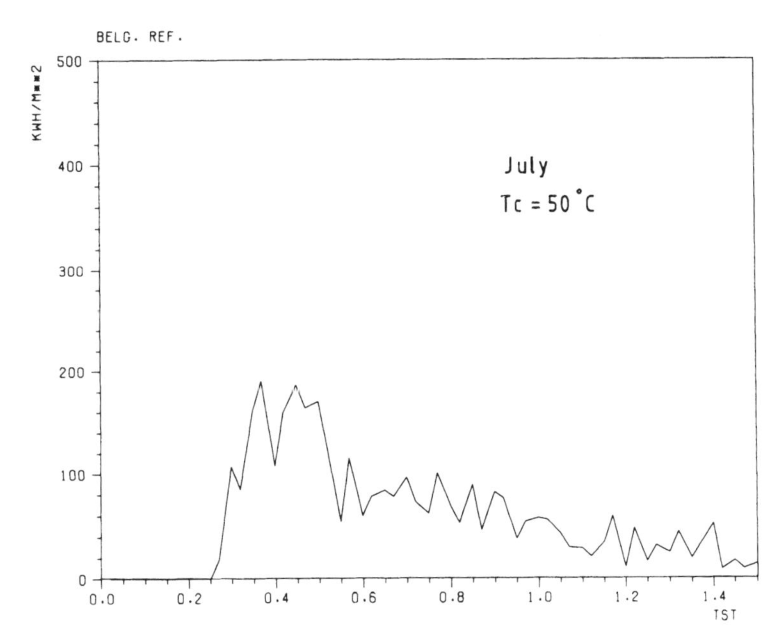

Fig. 10: Same as fig. 9 for July of the Belgian reference year

Table II. Computed energies of the Dru-collector for the Belgian Reference Year, resulting from a simulation calculation ("SIM") and from the RED-function method for a collector efficiency $\eta = 0.736-0.453\ T^*$.

MONTH	G	$T_{in} = 10^0$ C			$T_{in} = 30^0$ C			$T_{in} = 50^0$ C		
		SIM	RED	RED/SIM	SIM	RED	RED/SIM	SIM	RED	RED/SIM
1	27.5	14.9	15.1	1.01	7.2	7.2	1.00	3.4	3.4	1.00
2	47.7	27.4	28.1	1.03	15.8	15.9	1.01	8.5	9.2	1.08
3	77.4	49.7	52.7	1.06	31.6	32.4	1.03	19.1	20.1	1.05
4	92.6	61.2	66.3	1.08	40.7	43.1	1.06	24.7	27.0	1.09
5	115	78.8	85.9	1.09	52.4	56.4	1.08	30.8	35.1	1.14
6	136	101	106	1.05	72.0	77.9	1.08	48.0	53.1	1.11
7	111	82.4	87.3	1.06	53.4	58.0	1.09	30.5	35.0	1.15
8	113	86.1	89.8	1.04	59.1	63.4	1.07	36.9	40.4	1.09
9	113	86.3	88.6	1.03	63.7	65.7	1.03	44.0	45.3	1.03
10	57.8	40.7	42.9	1.05	24.6	25.2	1.02	15.1	15.9	1.05
11	28.9	17.1	17.7	1.04	8.5	8.5	1.00	4.2	4.5	1.07
12	19.8	9.1	9.2	1.01	4.3	4.3	1.00	1.6	1.8	1.13
YEAR	940	655	689	1.05	433	458	1.06	267	291	1.09

3.3.2 Test of the applicability of the Cumulative Frequency Curves

The precision of heat output calculations, performed on the basis of the cumulative frequency curves (CFC) has been determined by ECM for the Dru-collector for a period of 6 days with the data measured at Paris (see chapter 3.2). The results are summarized in table III. The precision of the result is essentially determined by an appropriate choice of the temperature of the ambient air. In method 1 the correct air temperature for each interval of 1 hour is used in the application of the HWB-formula. In method 2 the mean air temperature for the whole period is employed. For method 3 the mean air temperature from 9-15 h is used. In method 4 mean air temperatures were taken, which were calculated for each irradiation class. We observe an important improvement, when instead of the 24 h means, more sophisticated procedures are applied for the determination of the average temperature.

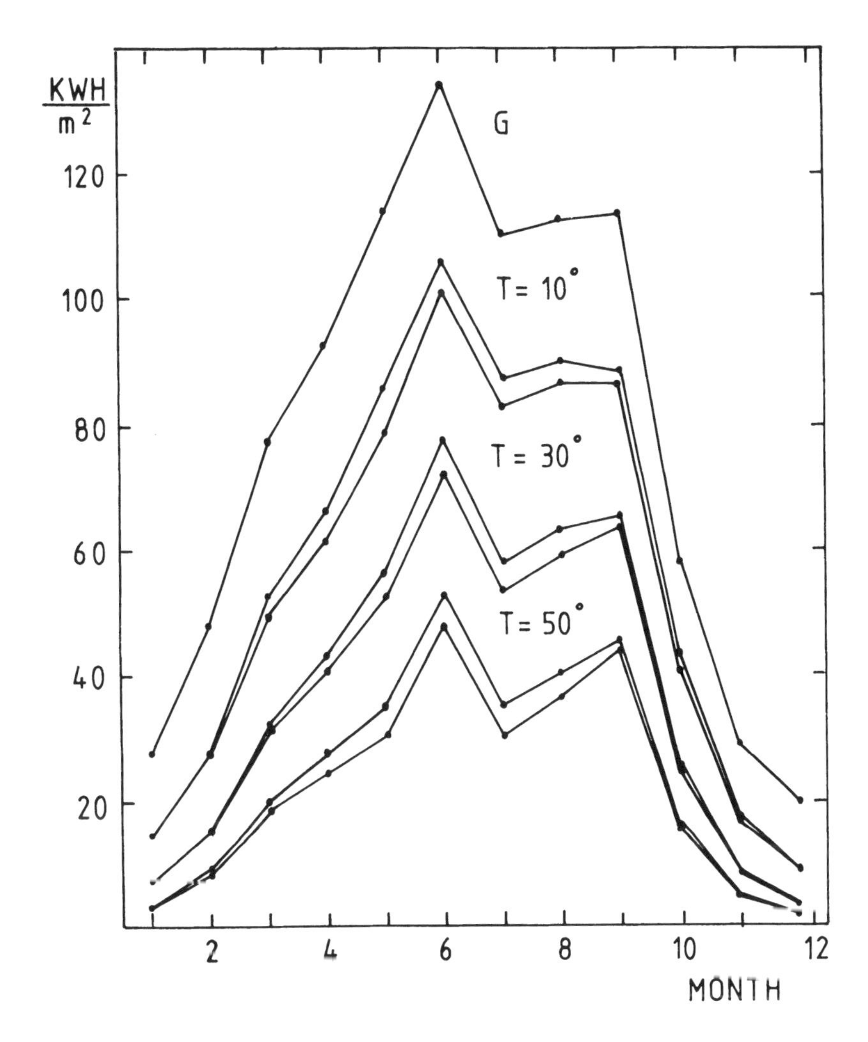

Fig. 11: Monthly heat output of the Dru-collector (slope = 51°) during the Belgian Reference Year. For every inlet temperature the upper curve is the result of the RED-function method, the lower was obtained with the help of the simulation programme COL2 (Faculté Polytechnique de Mons).

Table III. Comparison between several simplified methods calculating the useful energy output from the Dru-collector. The absolute energies (in KWH) and the deviation from the reference energy, obtained with the Minersol programme are given.

	Ti = 10 °C		Ti = 30 °C		Ti = 50 °C	
Reference energy	13.2		6.05		3.18	
Method 1 HWB - 1 hour	13.3	0,8 %	6.06	0,1 %	3.06	-3,8 %
Method 2 CFC, Ta	13.2	0,5 %	5.449	-10 %	2.57	-19 %
Method 3 CFC, Ta (9-15 h)	14.7	11 %	5.80	-4,2 %	2.75	-13 %
Method 4 - CFC, Ta for each class of irradiation	13.2	0,3 %	6.02	-0.4 %	3.04	-4,5 %

4. Conclusion

The deviations between the collector energies for the Belgian Reference Year, derived from the RED-function and through the simulation calculations with COL2, are due to effects, inherent in the method itself, but eventually also to an imperfect correspondence of the basic physical parameters, which were used in the simulation calculation, with the efficiency curve $\eta(T^*)$. But nevertheless, we conclude that for fixed temperatures the RED-function method yields in the average for flat plate collectors energies with good precision (deviation <10 %). Thus the main task, fixed in the contract of the KFA, is solved.

The question of time steps of data, used for simulation or for complex methods of data reduction, was studied with the help of relatively short data sets. Effects of some percent are observed, depending on the weather conditions. The Cumulative Radiation Frequencies Method was applied to a short data set, with encouraging results. Final conclusions can be drawn, when periods of 1 year are studied, including the question of appropriate mean ambient air temperatures.

References

1) Results and analysis of IEA round robin testing. IEA-report task III, KFA Jülich (1979)

2) Results and analysis of the round robin testing of the fourth solar collector in the European community programme. Edited by A. Derrick, University College, Cardiff, U.K. (1980)

3) Utilisation de courbes de fréquences cumulées d'irradiation globale pour le calcul des installations solaires. J. Adnot, B. Bourges, D. Campana, R. Gicquel Report of Ecole de Mines de Paris (1978)

4) Reduced meteorological data for the computation of the energy output of solar collectors.
H.R. Koch, progress report action 3.4, Brussels, April 1981

5) The layout of solar hot water systems, using statistical meteo- and heat demand data.
P. Kesselring, A. Duppenthaler
Contribution to the ISES 1979 International Congress on Solar Energy, May 28-June 1, Atlanta, Georgia, USA

ACTION 4

NEW MEASUREMENTS AND TECHNOLOGIES

Action 4.1 - Improvement of measurements in national networks - global, direct, diffuse irradiance and irradiance on inclined surfaces
Action leader : J.W. GRUETER, Kerforschungsanlage Jülich, Federal Republic of Germany

Action 4.2 - Special measurements
Action leader : J.-L. PLAZY, Commissariat à l'Energie Solaire, Sophia Antipolis, Valbonne, France

Action 4.3 - Satellite image processing
Action leader : R.M. POISSON, Centre de Télédétection et d'Analyse des Milieux naturels, Sophia Antipolis, Valbonne, France

ACTION 4.1 - IMPROVEMENT OF MEASUREMENTS IN NATIONAL NETWORKS GLOBAL, DIRECT, DIFFUSE IRRADIANCE AND IRRADIANCE ON INCLINED SURFACES

Action leader : J.W. GRüTER
Kernforschungsanlage Jülich GmbH
STE/Postfach 1913
D - 5170 JUELICH

Participants :

Met DK	Meteorologisk Institut Copenhagen (Frydendahl)
DWD	Meteorologisches Observatorium Hamburg, (Dehne)
Met GB	Meteorological Office, Brackwell (Durbin)
Met GR	Meteorological Institute, National Observatory, Athens (Lalas)
IFA	Istituto de fisica dell'atmosfera, Rome (Guerrini)
Met IR	Valentia Observatory, Cahirceveen (Murphy, McWilliams)
Met NL	Koninklijk Nederlands Meteorologisch Instituut, De Belt (Slob)

Tasks :

- Improvement of instrumentation (in particular for diffuse radiation)
- Densification of networks for global, direct and diffuse irradiance
- Measurements on inclined surfaces

1. ABSTRACT

Action 4.1, part of Project F of the Solar Energy R&D programme, deals with the radiation networks of Member States. New equipment and measuring techniques for networks are assessed and eventually installed by the national services, all of which participate in this action. The current programme supports the meteorological services of four Member States in upgrading their networks for global, diffuse and direct radiation. Progress continues at different rates in these countries. The measuring technique for diffuse radiation with shadow bands is assessed. A correction formula, generally applicable, permits improvement of the accuracy of this kind of measurement up to $\pm$ 20 W/m^2. One and a half year measured data of radiation measurements on inclined surfaces have been collected in Cabauw, the Netherlands, and are available on request. Data processing is in progress. "Solar Weeks" were organized for 1 - 14 November 1981, for the collection of meteorological data : global irradiation on a horizontal plane, diffuse global irradiation on a horizontal plane, sunshine duration - all three as hourly sums - air temperature and wind speed as available on magnetic tape, cloud cover on paper print. It is expected that data will be available three months later.
A description of the radiation networks is envisaged by the action leader. First data were collected. A draft report will be available in 6 months' time.

2. GENERAL SURVEY

Radiation mapping, like the European Solar Radiation Atlas, is based on measurements obtained from radiation networks. These networks are maintained by national meteorological services. During the past 3 years the services have been encouraged by the Commission to upgrade and densify their networks for global, diffuse and also to some extent direct radiation. Denmark, Italy, Ireland and the United Kingdom made substantial efforts, supported by the Commission. In the Federal Republic of Germany the number of stations was increased by a combined project with the German research minister, who financed the installation, while the scientific research on diffuse radiation measurements was supported by the Commission.

Major research topics of action 4.1 in the current programme are :
- the improvement of measuring techniques for diffuse radiation;
- spatial correlation studies in radiation, with Denmark as an example, which now has a fairly dense network.

A summary of the work of each participant is given at the end of this report.

Since the contractors' meeting in April 1981 one action meeting has been held in Athens, from 10 to 12 November 1981, together witch Action 1.

Progress in network instrumentation

In Italy only 9 of the 30 projected stations are running with the new equipment, and 6 stations will be added by the end of 1981. The participant is confronted by formidable administrative problems which have caused delays. In Denmark all 19 new stations are now running but there are still problems with the electronic equipment which leads to a relatively poor reliability of the stations. The new instruments of the Irish network are in place. Densification of the British network is in progress.

Diffuse sky radiation

Diffuse sky radiation can now be measured by shadow band techniques with an accuracy of $\pm$ 20 W/m^2, as the development of correction functions has reached a high level of accuracy.

The major work has been done in Hamburg. The contribution of the Irish participant could not be assessed so far owing to lack of discussions. The status of the Hamburg study is described below.

Operation of the pyrheliometers of type NIP (Eppley) and type Linke-Feussner - Actinometer (Kipp & Zonen) as well as the shade disk pyranometer (shaded view angle : 9.6°) has been continued at Hamburg.

To derive a new correction formula for the Hamburg shade ring both the hourly sums of diffuse solar radiation D_{mD} obtained from the shade disk pyranometer and the corresponding values of D_c, calculated by

$D_c = G - I \sin\gamma$ (I : direct solar radiation; γ : elevation of sun; G : global radiation) have been used as reference values. Based on hourly sums of D_{mR} (Hamburg shade ring) and G which were measured applying pyranometers of type CM 10 the following correction formula has been computed :

$$f = A+B(D_{mR}/G)^3 + C.\delta+/D/\tau' \text{ with : } \delta = \text{declination of sun and}$$

$$\tau' = \ln [I_o \sin\gamma/(G-D_{mR})]$$

For the data set of the period January 1981 - May 1981 and the two reference values the resulting constants of the formula are listed below :

	Reference	A	B	C	D	Corr. -c.
f_I	D_c (I form NIP)	1.2486	-.1927	-.00066	-.06007	> .80)
f_{II}	D_{mD}	1.1623	-.1075	.00146	-.03108	> .78)

Generally f_I exceeds f_{II} by about 1 - 2 %, as expected according to the relatively low view angle of the NIP (5 - 7°) compared with the disk shaded angle.

A simple formula for transferring the correction factors of the Hamburg ring (b/r = .169) to ring devices with a different b/r ratio has been developed.

Theoretical correction formulas for the Hamburg ring have been derived by assumption of ideal spatial distributions of sky radiance. To measure real distributions of sky radiance the installation of a sky scanner with broadband fibre optics and broadband sensors has been started.

Future work

The correction formula f_{II} will be improved by

a) using the total set of proved data of 1981;
b) further testing the influence on the correlation coefficient of other functions and parameters;

c) modifying the functions employed with regard to possible physical relationships.

For all types of cloudiness sky radiances will be obtained from the sky scanner.

First steps will be taken to compare experimentally the Hamburg shade disk pyranometer with shade ring pyranometers used in EC networks.

At the Athens meeting it was proposed to assess the influence of parameters B...D on the accuracy of the correction.

Radiation on inclined surfaces

This work is carried out by the KNMI at Cabauw, The Netherlands.

Status

Measurements of solar irradiance on inclined surfaces ceased on 15 October 1981. At present, the available data are being processed from 6-minute values to hourly, daily and monthly totals. Evaluation of the processed data is then carried out, and special attention is paid to the interpretation of specific measurements.

Results

Data collection of the global irradiance on inclined surfaces and several special measurements has been carried out from March 1979 to October 1981.

Special attention is paid to the following subjects :
- the shadow-band correction factor;
- the influence of a 0.53 μm filter;
- the comparison of three Eppley (NIP) pyrheliometers;
- the comparison of the Eppley PSP with the Kipp CM10 pyranometer.

The results of this data evaluation procedure will be reported at the end of 1981 and future work will be determined from the results of the evaluation.

The raw data are already available to some extent on request.

Action coordination

The focus of the action coordination is directed to two topics
- a coherent description of the radiation networks in the EC in order to obtain a "data source catalogue" for new data;
- production of a data tape from a "solar week" for spatial correlative studies and/or as ground reference data for the satellite image processing methods.

Data source catalogue

During the last contractors' meeting in Brussels, April 1981, the action leader distributed a proposal of a format to describe the network stations. No major objections were forwarded to the action leader. Therefore he will go ahead to start the collection of station descriptions from the meteorological services. Other sources of information which will be synthesized into the catalogue are the IEA data source catalogue, the IEA handbook for radiation networks and the last survey of the WMO.

A first survey of the methods of data handling, maintenance routines, data checks was obtained at the contractors' meeting in Athens. A draft report will be distributed amongst the participants for corrections by January and forwarded to the next action meeting for approval.

While discussing the work of the Danish participant during the Athens meeting, emphasized by Mr. STEEMERS implications, it was felt that more effort should be spent on the assessment of the spatial distribution of radiation stations. A proposal for a concerted study within the action was discussed and approved in principle by the participants which the action leader should coordinate.

This proposal will be forwarded to the expert group.

This assessment study would be part of the catalogue which may contain the following topics :

- description of stations;
- description of instruments;
- description of maintenance, data checks, calibration;
- assessment of the spatial distribution of stations.

DATA

A survey of the data handling in the 9 different services was made at the Athens meeting. A description will be made available by the action leader by the end of 1981 and distributed among the participants. The following topics were discussed :

- o electronic equipment
  - make
  - price
  - structure
  - mass storage
  - scanning frequencies
  - maintenance
  - security
- o quality control
  - methods
- o data storage
- o combination of radiometric with climatic data
- o how can a user obtain the relevant raw data of a certain station
  - costs
  - address
  - delay after acquisition.

"Solar weeks"

The collection of data for a fortnight was started. The time interval was foreseen as 1 November - 14 November 1981. The parameters, as available, G, D, SS, T_{air}, V as hourly values on magnetic tape and cloud cover as paperprint will be collected from all stations in the Community by the action leader from the respective organisations. The collection will take at least 3 months. The action leader will report to the next meeting.

G = global radiation, D = diffuse radiation, SS = sunshine duration, T_{air} = air temperature, V = wind speed at 10 m height.

Various

At the Mexico meeting CIMO accepted the threshold value of 120 W/m^2 for all automatic sunshine recorders. The recommendation of this value has to be confirmed by the Congress of the WMO.

References

1. Solar radiation data source catalogue
Weine Josefsson, Marie-Louise Westerberg
The Swedish Meteorological and Hydrological Inst., SMHI.
Norrköping, Sweden, Oct. 1980.
This report is part of the work within the IEA Solar Heating and Cooling Programme, Task V : Use of existing meteorological information for solar energy application.

2. J.R. Latimer and T.K. Won
Recommendations concerning meteorological networks for solar energy applications.

3. R. Dogniaux, chairman of the Working group on radiation, World Meteorological Organization, Inventory of radiation measurements, Region VI (Europe), May 1978.

3. SHORT DESCRIPTION OF INDIVIDUAL CONTRACTORS' WORK

Name, organisation, country	Title and aims of work
Radiation networks	
M. Fryedahl/Dan.Met.Inst./Denmark responsible scientist : Mr. Allerup	Maintenance and data production from the Danish (radiation) network Contribution and improvement of measurement of G and D on horizontal surfaces at 20 stations, G on inclined surfaces at 2 stations and broadband short-wave

	measurements at one station. Studies on the spatial correlation of global radiation in Denmark.
Mr. Lalas, Mr. Karalis, Athens/ Univ., Meteoroligal Inst.	Installation of a primary radiation station at the National Observatory Athens Contract in negotiation.
Mr. Guerrini/IFA/Italy Mr. Lavagnini	Development of a network for direct measurement of solar radiation Continuation and installation and improvement of measurement of G and D at 30 stations in Italy. Data elaboration of sunshine, global radiation, air temperature and wind
Mr. Murphy/Ir.Met.Serv./Ireland	Extension of network of G and D stations and improvement of diffuse radiation measurement Installation of instrumentation to measure G and D at 2 new stations, Clones and Malin Head. Logging-electronic, hourly values.
Mr. Durbin/UK Met.Serv./UK	Extension of the observation of solar radiation over the UK Installation of instrumentation to measure G and D at 4 new stations Stornoway, Shanwell, Camborne, Hembsby and to measure I at Eskdalemuir.

Diffuse radiation

Mr. Kasten/DWD/F.R.G. reponsible scientist : Mr. Dehne	Improvement of measurment of diffuse sky radiation Corrections to be related to cloudiness, turbidity, spatial distribution of diffuse radiation.

Figure 1: New equipment for radiation networks of the meteorological services in the European Community

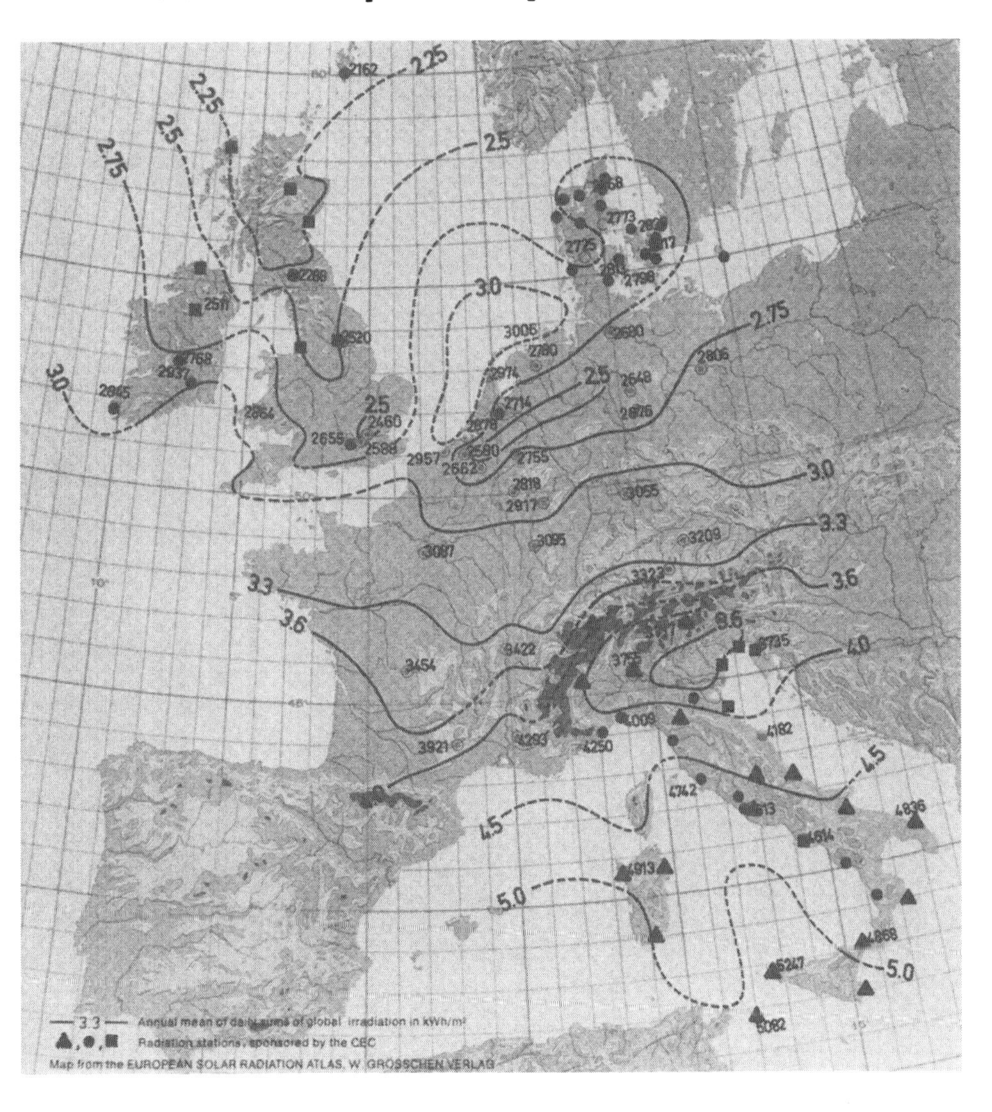

map of radiation stations, affected by Project F

- ⊙ stations used for the maps of the European solar radiation atlas (already more than ten years of data)
- ● stations running with improved equipment
- ■ stations newly equipped until Dec.1981
- ▲ stations foreseen for upgrading in the next two years.

Figure 2: spread of calibration factors of pyranometer sub-standards in use by the meteorological services in the European Community determined summer 1981 in Carpentras, France
from: Comparaison des pyranomètres étalons secondaires
J.L.Plazy, action 1,1981

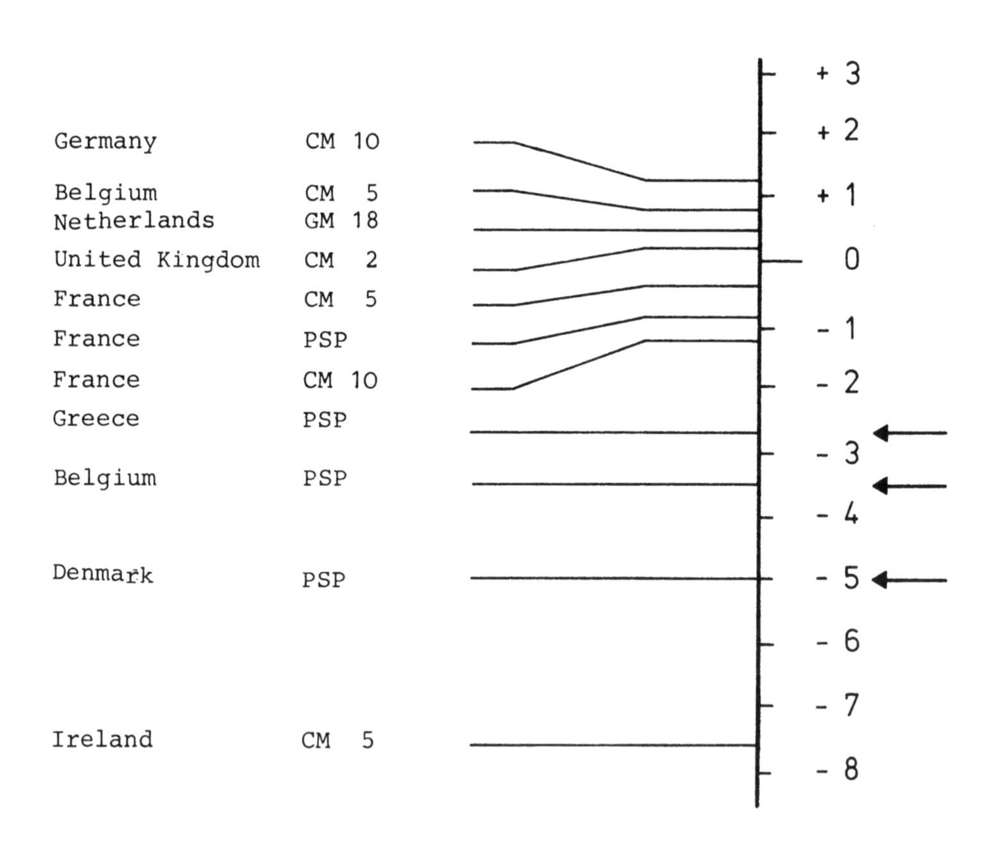

Italy has not participated

m: manufacturers' calibration
all other instruments are calibrated by the met.services

Figure 3: Diffuse sky radiation
Definition of correction factors f, k, k_{geo} for shade rings

$$f = k\,k_{geo} = \frac{G' - I'\sin\gamma}{D'_{SR}}$$

G' = global
D' = diffuse — solar irradiance
I' = direct normal incidence
γ = solar elevation
D'_{SR} = measured irradiance by a pyranometer shaded by a ring of radius r, width b

$$k_{geo} = \frac{1}{1-x} \text{ with } x = (2b/\pi r)\cos^3\delta(\sin\phi\,\sin\delta\cdot t_o + \cos\phi\,\cos\delta\,\sin t_o)$$

δ = solar declination, ϕ = latitude of the station
$\cos t_o = -\tan\phi\,\tan\delta$

k to be determined
= 1 if the sky radiance is isotropic

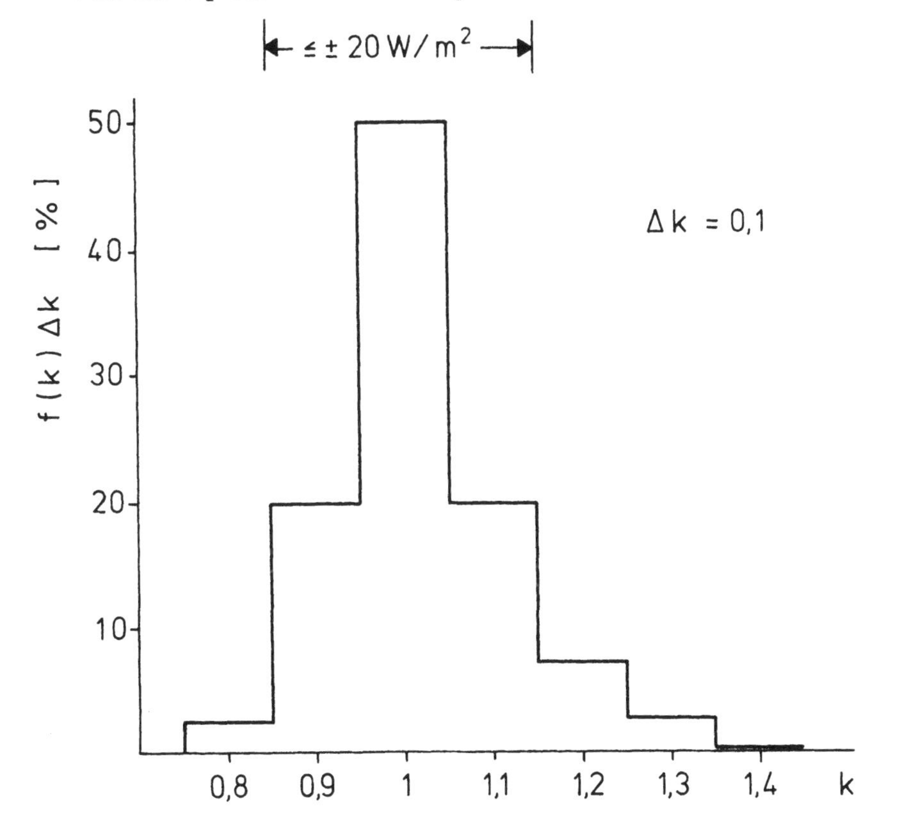

graph: frequency curve of correction factor k, measured winter 1980/81 in Valentia, Ireland

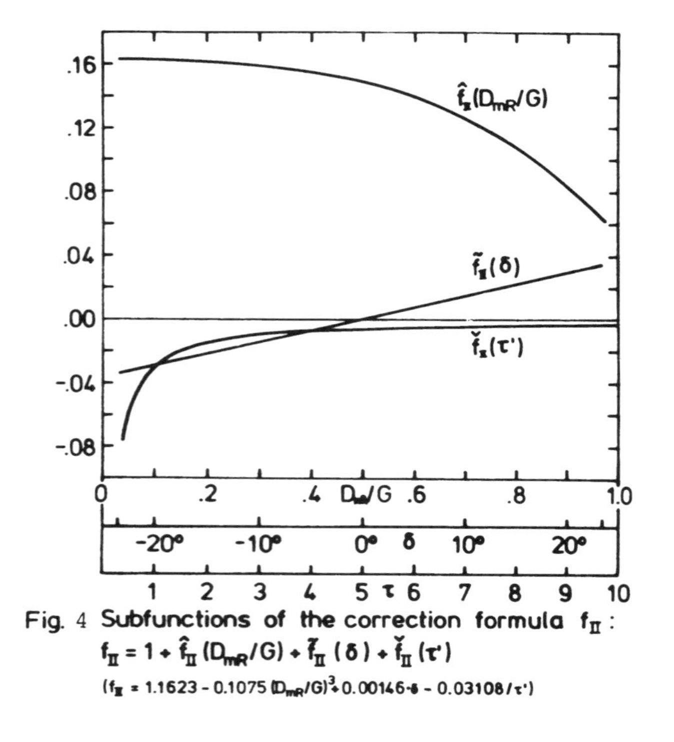

Fig. 4 Subfunctions of the correction formula f_{II}:
$f_{II} = 1 + \hat{f}_{II}(D_{mR}/G) + \tilde{f}_{II}(\delta) + \check{f}_{II}(\tau')$
$(f_{II} = 1.1623 - 0.1075\,(D_{mR}/G)^3 + 0.00146\cdot\delta - 0.03108/\tau')$

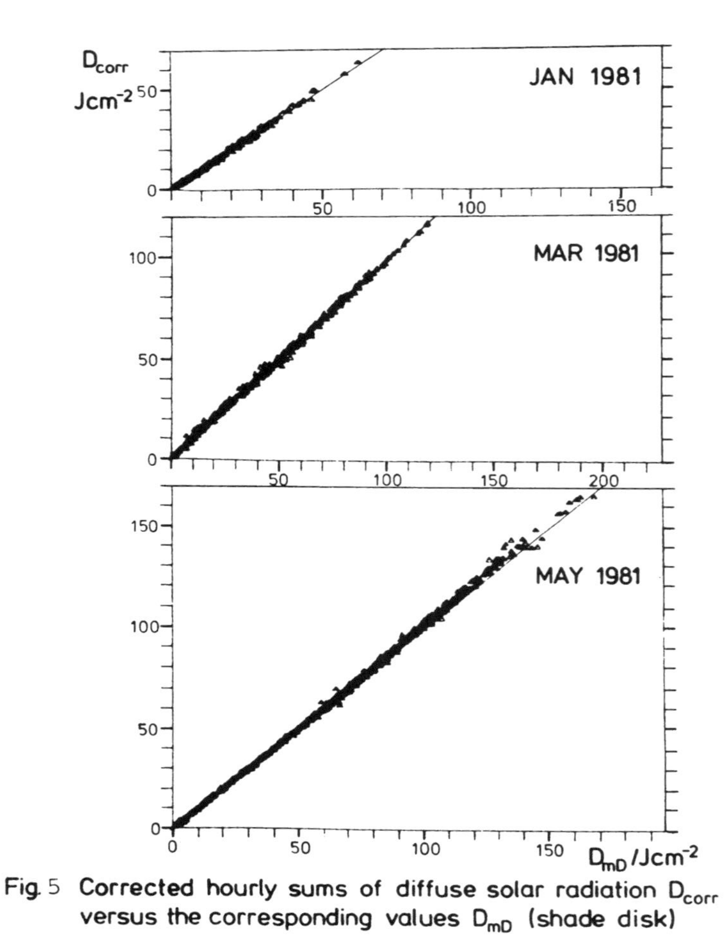

Fig. 5 Corrected hourly sums of diffuse solar radiation D_{corr} versus the corresponding values D_{mD} (shade disk)

Mr. McWilliams/Val./Ireland	Analysis of existing G and I measurements at Valentia to give accurate D for comparison with measurements of D with shading ring. Corrections related to cloudiness, D/G etc.

Global radiation on inclined surfaces

Mr. Slob / KNMI/Netherlands	Solar radiation measurement on inclined and horizontal surfaces Continuation and improvement of measurement of global radiation on a horizontal and a range of inclined surfaces (both 0.3 - 3 μm and 0.5 - 5 μm), diffuse radiation with a shading ring, short-wave albedo. Plans to commence measurement of downcoming long-wave radiation and use of sky camera to aid with analysis of inclined surface measurement. Logging 6 min averages. All measurements made at Cabauw/Netherlands.

EXPLANATION OF ILLUSTRATIONS

Radiometric networks

Figure 1 contains a map of radiation stations in the EC affected by Project F either by using ten-year series of data for the production of radiation maps (Action 3.1) or by sponsoring the improvement of their equipment.

Figure 2 shows the spread of calibration factors of pyranometer substandards for different countries and different makes. Obviously all substandards but one, so long as they are calibrated by the services themselves, show deviations of less than 1.5 %. The reason for the large deviation of the Irish instrument is not understood. Examination of the instrument by the Irish Met. Service is in progress.

One may conclude that the calibrations of the different networks in the Community are in good agreement, as already determined by the international intercomparison of the primary standards, the pyrheliometers held in Carpentras 1978.

A new intercomparison of the same instruments for winter conditions will be held in Carpentras, France, in Jan. 1982.

Diffuse sky radiation

Figures 3, 4 and 5 deal with the problem of diffuse sky radiation measurements with shaded pyranometers.

Figure 3 shows the spread of relative deviations, measuring the diffuse sky radiation by a shaded pyranometer, from the true values as determined from G and I, the global and direct normal incidence radiation. The respective definitions are given and a correction factor for the sky cut off by the shade ring, assuming isotropic distribution of the sky radiance. This correction factor is applied to the output of the shaded instrument before calculating the values for the graph on the figure by the Irish contractor with data of the Valentia Observatory.

Figure 4 shows the contributions of the different parts of the correction formula for factor f, developed by the Hamburg group, as defined in Figure 3 and described in the text.

In Figure 5 the values of the diffuse sky radiation as determined by two methods are plotted in a correlation point plot. Each point represents a value point where the radiation is measured either by a pyranometer, shaded by a ring, applying the appropriate correction formula f or by a pyranometer shaded by a disc of about 10° shaded view angle around the sun.

The straight lines represent identical values determined by both methods.

The majority of the deviation are much less than 20 W/m^2.

ACTION 4.2 - SPECIAL MEASUREMENTS

Action leader : M. J.-L. PLAZY
Commissariat à l'Energie Solaire
Sophia-Antipolis
F - 06560 VALBONNE

Participants : Direction de la Météorologie PARIS — ESF-005 F
M. J.A. BEDEL

Institut für Meteorologie — ESF-015 D
Johannes Gutenberg Universität MAINZ
Prof. R. EIDEN

Institut Royal Météorologique de Belgique UCCLE — ESF-003 B
Prof. R. DOGNIAUX

Laboratoire d'Energétique Solaire — ESF-009 F
CNRS - ODEILLO
M. J.F. TRICAUD

Irish Meteorological Service DUBLIN — ESF-018 E
Mr. E.J. MURPHY

Tasks : Development of instrumentation for specific needs in solar energy.

Turbidity, solar aureole, spectral and long-wave radiation, distribution of the luminance.

I. INTRODUCTION

The aim of action 4.2 is to promote the development of instrumentation for specific needs in solar energy and particularly in the fields of terrestrial radiation, the spatial distribution of the sky luminance, circumsolar radiation, the turbidity and the solar spectrum after traversing the atmosphere. During the last programme some instruments were developed for turbidity, infra-red and spectral measurements and these have been poned in operation.

Because of the large dispersion of the topics of this action and the difficulties encountered by some of the contractors to obtain the necessary finance from their government, no action meeting has been held during the last six months.

2. DESCRIPTION OF THE WORK

2.1 Circumsolar sky radiation

The radiation flux of the direct solar beam is diminished by absorption and scattering processes causes by atmospheric molecules, aerosols and cloud particles. This attenuation is partly compensated by the diffuse radiation coming from the circumsolar part of the sky. For instance, a solar power plant using focusing optical systems may be equipped with low focusing, inexpensive systems to take into account the circumsolar radiation to reduce atmospheric extinction losses.

The Johannes Gutenberg University in Mainz (Federal Republic of Germany) has developed a special instrument for circumsolar radiation measurements. The angular distribution of the sky radiance is measured continuously from 1 to 10 degrees distance from the centre of the sun. The field of view of the photometer tube is 0.5° and the error is less than 1 % of the measured sky radiance. The detector system uses a silicon photovoltaic detector combined with a flat filter. It is calibrated in absolute radiometric terms over the spectral range of 350 to 1100 nm. The radiant flux density F depending on the angular distance from the centre of the sun is obtained by integrating the primarily measured radiance :

$$F = 2 \int_{0.5}^{\varphi} B(\varphi) \sin\varphi \, d\varphi$$

In parallel, the Linke turbidity factor T (1) is being measured for five monochromatic wavelengths at 402, 451, 551.5, 652 and 696 nm.

This equipment has been checked in Mainz by comparison with theoretical values and was installed in Adrano (Sicily) in November 1981 in order to compare the efficiency at the solar power plant according to the circum-solar sky radiation.

2.2 Atmospheric turbidity

A climatological study of the atmospheric turbidity is being prepared by the Direction de la Météorologie (France) using the direct solar radiation measurements on a six-minute basis recorded in Trappes and Carpentras. A special file on magnetic tape has been constituted, and the climatological study can now be started.

2.3 Spectral measurements

Spectral measurements of global solar radiation have been systematically and continuously recorded during the whole year 1980 at Uccle (Belgium) by the Institut Royal Météorologique de Belgique. These records are obtained by means of a measuring equipment composed of an eight-head pyranometer operating at eight wavelengths : 315, 400, 446, 545, 646, 730, 816 and 914 nm. The calibration and the periodic checks are carried out with a double monochromator type Cary 14, and by a method based on the use of an integrating sphere as radiation source.

2.4 Long-wave infra-red incoming radiation

The long-wave incoming radiation is now measured systematically at Trappes and Carpentras on hourly amounts with pyrradiometers developed during the first R&D solar programme of the European Communities. A special file of hourly data has been constituted and will be statistically analysed during the next month. The spectral range of the measured radiation is 0.3 to

50 micrometres and the long-wave radiation is deduced from it by substracting the global solar radiation measured with a pyranometer.

2.5 Reflected and direct radiation measurements, Albedo

Special equipment or measuring albedo, direct radiation, reflected radiation on inclined surfaces, and sunshine duration have been installed in 1981 mainly at the Valentia Observatory (Eire) and Odeillo (France) to complete the equipment of these main radiometric stations.

In Odeillo the reflected radiation on a vertical plane is measured with a pyranometer on which is placed a quarter of a sphere to prevent the radiation coming from the sun or the sky reaching the thermopile. The sunshine duration is deduced from the direct radiation measurement made with a pyrheliometer by counting the time during which the direct radiation is above five levels : 110, 200, 500, 600 and 800 W m^{-2}. The time step of the sampling period is 12 seconds. This kind of data will make it possible to deduce the operating duration of a solar energy system taking into account the threshold for operation.

In Valentia the diffuse radiation measurements will be used to calculate the errors caused by using a shading ring which cut off a part of the diffuse radiation coming from the sky.

3. RESULTS

At the present time, results are as follows :

- constitution of a special file of hourly total incoming radiation for Carpentras (5 years) and Trappes (18 months);
- constitution of a special file of six-minute and hourly values of the direct radiation for Carpentras and Trappes (2 years). Constitution of this file has taken longer than initially projected because of some missing data, owing to some problems with the data acquisition system;

- direct, diffuse and reflected radiation measurements are now made at Valentia, and at Odeillo from the first days of October. In addition at Odeillo, the duration of overstepping five levels of direct solar radiation (110, 200, 500, 600 and 800 W m^{-2}) is recorded.

An example of these results, giving the sunshine duration obtained with a pyrheliometer and a threshold of 110 W m^{-2} on an hourly basis is given in Figure 1.

- a methodology for the calibration of spectral measurement instruments using an integrating sphere and a monochromator;

- a full year of spectral data at Uccle (1980).

  These data are now published by the IRMB. Misc Serie B n° 52 and 53 in 1981 under the title "Distribution spectrale du rayonnement solaire à Uccle, 1er semestre 1980 et 2ème semestre 1980".

  The results are presented in the form of daily tables giving respectively the spectral irradiance (half-hourly variations) and the spectral radiant exposures (half-hourly and daily sums). Some examples are given in Figure 2.

- a reliable instrument for circumsolar radiance and irradiance measurements. A convenient system to process the data is being developed and is now almost finished.

  An example of the data obtained with such an instrument is given in Figure 3, where it is possible to see the radiance B and the irradiance F for various sky conditions according to the angular distance from the sun.

	Date	Air mass	Turbidity	Radiance 1	Radiance 10	Irradiance 1	Irradiance 10
CLEAR SKY	4/10/81	1.75	6	1169.0	323.3	0.94	40.6
	4/24.81	1.47	2	298.4	87.0	0.24	10.3
CIRRUS	4/15.81	1.5	-	7707.0	298.4	7.7	73.0
	4/24/81	1.58	-	1902.1	126.8	2.9	23.8
				$W\ m^{-2}$	sr^{-2}	$W\ m^{-2}$	

The results show that under some conditions the irradiance produced by a 10-degree circle around the sun disc can reach 10 % of the direct radiation.

It would be interesting to give at the same time the value of the direct radiation coming from the sun disc alone and the direct radiation measured with a pyrheliometer which has a non-negligible opening angle. In fact, most of the commonly used pyrheliometers have an opening angle of about 5 degrees.

On the two graphs (Figs. 4 and 5) it is possible to assimilate the variation of the circumsolar radiance and irradiance in relation with the angular distance to a straight line on logarithmic axes. This should simplify the computing models which could be deduced from this work to calculate directly the circumsolar irradiance using only the turbidity factor or the type of clouds.

4. CONCLUSIONS

All the equipments are now operating and data collection has begun in all cases.

The products will be delivered on time or with a slight delay.

The study of the relation between direct solar radiation and turbidity will start at the beginning of 1982. The first results will be presented during the third quarter of 1982. The climatological study on I.R. radiation has began on 1 November, and will be finished in June 1982 for the main results.

The reports on spectral radiation can be considered as finished. The Uccle measurements have been stopped and it seems difficult to reactivate them because of administrative problems with the Belgian Government.

The equipment for measuring circumsolar radiation will be transferred in some weeks to Adrano near the Eurhelios installation in Sicily.

A consolidated report on these special measurements will be produced by the action leader 4.2 for June 1982. A draft will be proposed within 6 months giving a description of the equipment used for these special purposes.

ODEILLO — RELEVE MENSUEL D'INSOLATION AVEC SEUIL DE 110W/M2

		INSOLATION	
LATITUDE : 42 29' N	PYRHELIOMETRE EPPLEY N.I.P. NO:19054E6	HELIOGRAPHE: PYRHELIOMETRE	ANNEE: 1981
LONGITUDE: 02 07' E	AVEC SEUIL: 110W/M2	AVEC SEUIL: 110W/M2	
ALTITUDE : 1580 M	INTEGRATEUR HP85	UNITE EMPLOYEE: 1/10 HEURE	MOIS: OCTOBRE
	UNITE EMPLOYEE: 1/10 HEURE		

JOUR	0-1	1-2	2-3	3-4	4-5	5-6	6-7	7-8	8-9	-10	-11	-12	MAT	JOUR	-13	-14	-15	-16	-17	-18	-19	-20	-21	-22	-23	-24	SOIR	TOT	M	S	T
274	0	0	0	0	0	0	0	0	0	0	0	0	0	1	0	0	6	0	0	0	0	0	0	0	0	0	8	8	0	8	8
275	0	0	0	0	0	0	1	0	1	3	3	1	11	2	0	1	4	0	9	0	0	0	0	0	0	0	16	28	11	16	28
276	0	0	0	0	0	0	0	2	4	5	1	0	14	3	0	0	0	1	0	0	0	0	0	0	0	0	2	16	14	2	16
277	0	0	0	0	0	0	0	0	0	0	7	7	15	4	9	9	10	10	10	8	0	0	0	0	0	0	58	73	15	58	73
278	0	0	0	0	0	0	0	0	5	9	10	9	34	5	10	10	10	10	10	1	0	0	0	0	0	0	51	85	34	51	85
279	0	0	0	0	0	0	4	10	10	9	9	10	52	6	10	10	10	10	10	4	0	0	0	0	0	0	54	106	52	54	106
280	0	0	0	0	0	0	3	6	8	10	10	10	48	7	10	10	9	10	10	4	0	0	0	0	0	0	53	102	48	53	102
281	0	0	0	0	0	0	0	10	10	10	10	10	50	8	10	10	6	0	4	0	0	0	0	0	0	0	30	81	50	30	81
282	0	0	0	0	0	0	0	0	1	0	1	8	11	9	1	0	0	0	5	0	0	0	0	0	0	0	7	18	11	7	18
283	0	0	0	0	0	0	1	0	2	1	8	9	22	10	4	1	6	7	4	1	0	0	0	0	0	0	25	48	22	25	48
TD	0		0		0		11		43		61		261		56		63		62		0		0		0		307		261	307	
		0		0		0		29		48		67				53		51		20		0		0		0		569			569
284	0	0	0	0	0	0	0	0	2	0	1	0	5	11	0	0	2	0	2	0	0	0	0	0	0	0	5	10	5	5	10
285	0	0	0	0	0	0	0	0	0	0	0	0	1	12	5	2	1	0	0	0	0	0	0	0	0	0	8	10	1	8	10
286	0	0	0	0	0	0	1	3	8	1	0	0	15	13	0	0	0	0	0	0	0	0	0	0	0	0	0	15	15	0	15
287	0	0	0	0	0	0	1	4	10	10	10	10	45	14	10	10	10	10	10	1	0	0	0	0	0	0	51	97	45	51	97
288	0	0	0	0	0	0	1	10	9	10	10	10	50	15	10	10	10	10	10	2	0	0	0	0	0	0	52	103	50	52	103
289	0	0	0	0	0	0	0	10	10	10	10	10	50	16	10	10	10	10	10	0	0	0	0	0	0	0	50	101	50	50	101
290	0	0	0	0	0	0	0	10	10	10	10	10	50	17	10	10	10	10	10	0	0	0	0	0	0	0	50	100	50	50	100
291	0	0	0	0	0	0	0	9	8	10	8	3	39	18	0	0	5	0	2	0	0	0	0	0	0	0	8	48	39	8	48
292	0	0	0	0	0	0	0	0	0	0	0	0	0	19	0	0	0	0	0	0	0	0	0	0	0	0	0	0	0	0	0
293	0	0	0	0	0	0	0	0	0	0	0	0	0	20	0	0	0	0	0	0	0	0	0	0	0	0	0	0	0	0	0
TD	0		0		0		6		59		50		259		45		49		44		0		0		0		229		259	229	
		0		0		0		46		51		44				43		40		6		0		0		0		488			488
294	0	0	0	0	0	0	0	0	0	0	0	0	0	21	0	0	0	0	0	0	0	0	0	0	0	0	0	0	0	0	0
295	0	0	0	0	0	0	0	0	0	0	0	0	0	22	0	0	0	0	0	0	0	0	0	0	0	0	0	0	0	0	0
296	0	0	0	0	0	0	0	0	0	0	0	0	0	23	0	0	0	0	0	0	0	0	0	0	0	0	0	0	0	0	0
297	0	0	0	0	0	0	0	0	0	0	0	0	0	24	0	0	0	0	0	0	0	0	0	0	0	0	0	0	0	0	0
298	0	0	0	0	0	0	0	0	0	0	0	0	0	25	0	0	0	0	0	0	0	0	0	0	0	0	0	0	0	0	0
299	0	0	0	0	0	0	0	0	0	0	0	0	0	26	0	0	0	0	0	0	0	0	0	0	0	0	0	0	0	0	0
300	0	0	0	0	0	0	0	0	0	0	0	0	0	27	0	0	0	0	0	0	0	0	0	0	0	0	0	0	0	0	0
301	0	0	0	0	0	0	0	0	0	0	0	0	0	28	0	0	0	0	0	0	0	0	0	0	0	0	0	0	0	0	0
302	0	0	0	0	0	0	0	0	0	0	0	0	0	29	0	0	0	0	0	0	0	0	0	0	0	0	0	0	0	0	0
303	0	0	0	0	0	0	0	0	0	0	0	0	0	30	0	0	0	0	0	0	0	0	0	0	0	0	0	0	0	0	0
304	0	0	0	0	0	0	0	0	0	0	0	0	0	31	0	0	0	0	0	0	0	0	0	0	0	0	0	0	0	0	0
TD	0		0		0		0		0		0		0		0		0		0		0		0		0		0		0	0	
		0		0		0		0		0		0				0		0		0		0		0		0		0			0
TM	0		0		0		18		102		111		520		102		112		107		0		0		0		537		520	537	
		0		0		0		75		100		111				96		92		27		0		0		0		1058			1058
MM	0	0	0	0	0	0	1	4	5	5	6	6	28	MM	5	5	6	5	5	1	0	0	0	0	0	0	29	58	28	29	58

Fig. 1

ECLAIREMENT ENERGETIQUE SPECTRAL SOLAIRE EN W m^{-2} nm^{-1}

SPECTRAL SOLAR IRRADIANCE IN W m^{-2} nm^{-1}

DATE : 80/ 8/ 1

TEMPS SOLAIRE	HAUT SOL MOY	MASSE D' AIR	INS REL	LONGUEJRS D'ONDE EN NM 315	400	446	545	646	730	816	914
3.30- 4.00											
4.00- 4.30				0,00	0,00	0,01	0,00	0,00	0,01	0,01	0,00
4.30- 5.00	2,65	16,526	0,0	0,00	0,02	0,03	0,03	0,03	0,03	0,02	0,02
5.00- 5.30	7,02	7,710	0,0	0,00	0,05	0,08	0,07	0,07	0,06	0,05	0,03
5.30- 6.00	11,55	4,878	0,0	0,00	0,05	0,07	0,06	0,05	0,05	0,04	0,03
6.00- 6.30	16,19	3,543	0,0	0,01	0,13	0,18	0,16	0,14	0,11	0,09	0,06
6.30- 7.00	20,90	2,782	0,2	0,02	0,22	0,30	0,29	0,26	0,20	0,17	0,11
7.00- 7.30	25,64	2,300	0,3	0,03	0,29	0,39	0,38	0,33	0,26	0,21	0,14
7.30- 8.00	30,35	1,972	0,8	0,05	0,50	0,70	0,72	0,64	0,51	0,42	0,29
8.00- 8.30	34,99	1,739	0,7	0,06	0,50	0,68	0,68	0,59	0,46	0,37	0,26
8.30- 9.00	39,48	1,570	0,5	0,07	0,55	0,78	0,78	0,68	0,54	0,44	0,30
9.00- 9.30	43,74	1,444	0,5	0,09	0,63	0,90	0,89	0,77	0,60	0,49	0,34
9.30-10.00	47,67	1,351	1,0	0,12	0,84	1,23	1,24	1,10	0,87	0,70	0,49
10.00-10.30	51,13	1,283	1,0	0,13	0,92	1,38	1,40	1,23	0,98	0,79	0,55
10.30-11.00	53,97	1,235	0,2	0,09	0,46	0,64	0,63	0,53	0,43	0,35	0,23
11.00-11.30	56,00	1,205	0,5	0,10	0,64	0,92	0,93	0,80	0,64	0,52	0,36
11.30-12.00	57,06	1,190	0,7	0,14	0,87	1,28	1,27	1,11	0,88	0,71	0,49
12.00-12.30	57,06	1,190	1,0	0,15	0,95	1,42	1,42	1,25	0,99	0,79	0,55
12.30-13.00	56,00	1,205	1,0	0,14	0,87	1,28	1,30	1,14	0,89	0,73	0,50
13.00-13.30	53,97	1,235	1,0	0,14	0,89	1,31	1,34	1,17	0,92	0,74	0,51
13.30-14.00	51,13	1,283	1,0	0,13	0,84	1,22	1,24	1,09	0,84	0,68	0,47
14.00-14.30	47,67	1,351	1,0	0,12	0,79	1,14	1,18	1,03	0,80	0,65	0,44
14.30-15.00	43,74	1,444	0,2	0,09	0,49	0,68	0,68	0,58	0,45	0,37	0,24
15.00-15.30	39,48	1,570	0,0	0,07	0,33	0,45	0,44	0,36	0,29	0,24	0,15
15.30-16.00	34,99	1,739	0,0	0,05	0,19	0,26	0,26	0,21	0,17	0,14	0,09
16.00-16.30	30,35	1,972	0,0	0,05	0,23	0,32	0,32	0,27	0,21	0,17	0,11
16.30-17.00	25,64	2,300	0,0	0,04	0,19	0,26	0,25	0,21	0,17	0,14	0,09
17.00-17.30	20,90	2,782	0,0	0,02	0,08	0,11	0,10	0,08	0,07	0,06	0,03
17.30-18.00	16,19	3,543	0,0	0,02	0,07	0,10	0,10	0,08	0,07	0,06	0,03
18.00-18.30	11,55	4,878	0,0	0,02	0,10	0,13	0,13	0,11	0,09	0,07	0,05
18.30-19.00	7,02	7,710	0,0	0,01	0,06	0,09	0,09	0,08	0,06	0,05	0,03
19.00-19.30	2,65	16,526	0,0	0,01	0,02	0,03	0,02	0,02	0,02	0,02	0,01
19.30-20.00				0,00	0,00	0,00	0,00	0,00	0,01	0,01	0,00
20.00-20.30											
20.30-21.00											

EXPOSITION ENERGETIQUE SPECTRALE SOLAIRE SEMI-HORAIRE ET JOURNALIERE EN J cm^{-2}

HALF HOURLY AND DAILY SPECTRAL SOLAR RADIANT EXPOSURE IN J cm^{-2}

DATE : 80/ 8/ 1

TEMPS SOLAIRE	HAUT SOL MOY	MASSE D' AIR	INS REL	400 G 300	500 G 400	600 G 500	700 G 600	800 G 700	900 G 800	1000 G 900	1000 G 300	GT	1000 G 300 /GT	DT/GT
3.30- 4.00														
4.00- 4.30				0,0	0,1	0,1	0,1	0,1	0,1	0,1	1	0		0,00
4.30- 5.00	2,65	16,526	0,0	0,2	0,6	0,6	0,6	0,5	0,4	0,3	3	4	0,76	1,00
5.00- 5.30	7,02	7,710	0,0	0,4	1,3	1,3	1,2	1,0	0,8	0,5	7	9	0,81	1,00
5.30- 6.00	11,55	4,878	0,0	0,4	1,1	1,1	0,9	0,8	0,6	0,4	5	6	0,88	1,00
6.00- 6.30	16,19	3,543	0,0	1,1	3,0	2,9	2,4	1,9	1,4	0,9	14	17	0,80	0,94
6.30- 7.00	20,90	2,782	0,2	1,8	5,1	5,2	4,5	3,6	2,7	1,7	25	31	0,79	0,87
7.00- 7.30	25,64	2,300	0,3	2,5	6,6	6,7	5,8	4,5	3,3	2,1	31	40	0,79	0,77
7.30- 8.00	30,35	1,972	0,8	4,3	11,9	12,7	11,3	[illegible]	6,7	4,3	60	77	0,78	0,55
8.00- 8.30	34,99	1,739	0,7	4,3	11,5	12,0	10,4	8,0	6,0	3,8	56	70	0,80	0,66
8.30- 9.00	39,48	1,570	0,5	4,9	13,1	13,8	12,0	9,3	7,0	4,6	65	79	0,82	0,67
9.00- 9.30	43,74	1,444	0,5	5,7	15,0	15,7	13,5	10,5	7,9	5,0	73	90	0,81	0,62
9.30-10.00	47,67	1,351	1,0	7,4	20,5	21,9	19,3	14,9	11,2	7,3	103	126	0,82	0,28
10.00-10.30	51,13	1,283	1,0	8,3	23,0	24,7	21,7	16,9	12,7	8,2	116	139	0,83	0,36
10.30-11.00	53,97	1,235	0,2	4,3	10,8	11,2	9,4	7,4	5,6	3,4	52	62	0,84	0,85
11.00-11.30	56,00	1,205	0,5	5,8	15,5	16,3	14,2	11,0	8,4	5,3	76	92	0,83	0,79
11.30-12.00	57,06	1,190	0,7	7,9	21,3	22,5	19,6	15,2	11,4	7,4	105	126	0,84	0,48
12.00-12.30	57,06	1,190	1,0	8,6	23,6	25,1	22,0	17,0	12,8	8,4	117	142	0,83	0,29
12.30-13.00	56,00	1,205	1,0	7,9	21,4	22,9	20,0	15,5	11,7	7,5	107	129	0,83	0,36
13.00-13.30	53,97	1,235	1,0	8,1	22,0	23,5	20,6	15,9	11,9	7,7	110	133	0,82	0,37
13.30-14.00	51,13	1,283	1,0	7,6	20,4	21,9	19,1	14,6	11,0	7,1	102	124	0,82	0,34
14.00-14.30	47,67	1,351	1,0	7,1	19,2	20,6	18,1	13,9	10,4	6,6	96	118	0,81	0,39
14.30-15.00	43,74	1,444	0,2	4,5	11,4	11,9	10,2	7,9	5,9	3,5	55	68	0,81	0,78
15.00-15.30	39,48	1,570	0,0	3,1	7,5	7,8	6,5	5,0	3,7	2,1	36	43	0,83	0,95
15.30-16.00	34,99	1,739	0,0	1,9	4,4	4,5	3,8	2,9	2,2	1,2	21	24	0,87	1,00
16.00-16.30	30,35	1,972	0,0	2,2	5,5	5,6	4,8	3,6	2,7	1,6	26	31	0,84	1,00
16.30-17.00	25,64	2,300	0,0	1,8	4,4	4,5	3,7	2,9	2,1	1,2	21	25	0,82	1,00
17.00-17.30	20,90	2,782	0,0	0,8	1,8	1,8	1,5	1,2	0,9	0,4	8	9	0,91	1,00
17.30-18.00	16,19	3,543	0,0	0,7	1,7	1,8	1,5	1,3	0,9	0,5	8	10	0,83	1,00
18.00-18.30	11,55	4,878	0,0	0,9	2,2	2,3	2,0	1,6	1,2	0,7	11	15	0,71	0,93
18.30-19.00	7,02	7,710	0,0	0,6	1,5	1,6	1,4	1,1	0,8	0,5	7	9	0,83	1,00
19.00-19.30	2,65	16,526	0,0	0,2	0,5	0,4	0,4	0,4	0,2	0,1	2	3	0,74	1,00
19.30-20.00				0,0	0,0	0,0	0,1	0,1	0,1		0	1	0,30	1,00
20.00-20.30														
20.30-21.00														
JOURNEE			0,4	115,2	307,9	324,8	282,5	219,0	164,7	104,3	1519	1851	0,82	0,54

Fig. 2

SPECTRAL CONCENTRATION OF SOLAR IRRADIANCE

DENSITE SPECTRALE D'ECLAIREMENT SOLAIRE

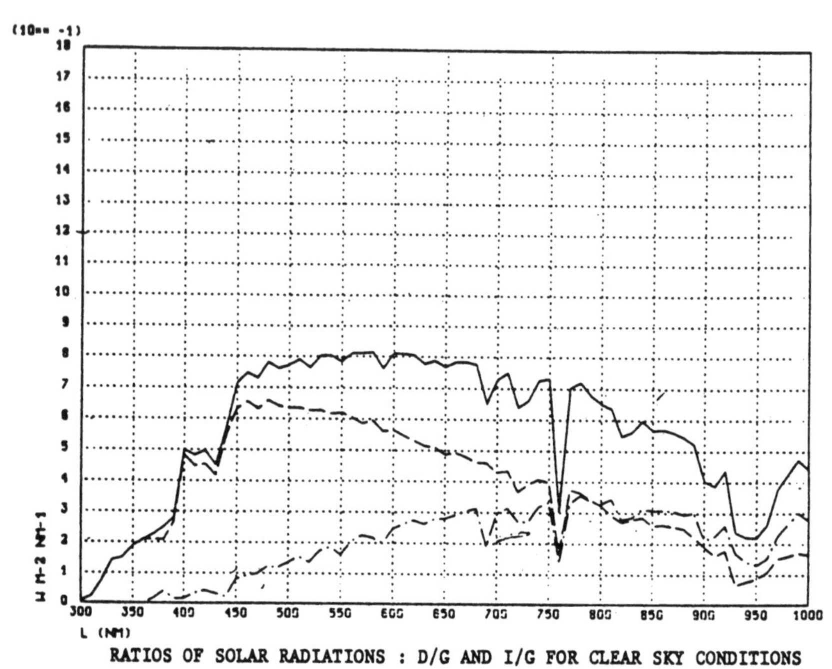

RATIOS OF SOLAR RADIATIONS : D/G AND I/G FOR CLEAR SKY CONDITIONS

RAPPORTS DES RADIATIONS SOLAIRES : D/G et I/G PAR CIEL SEREIN

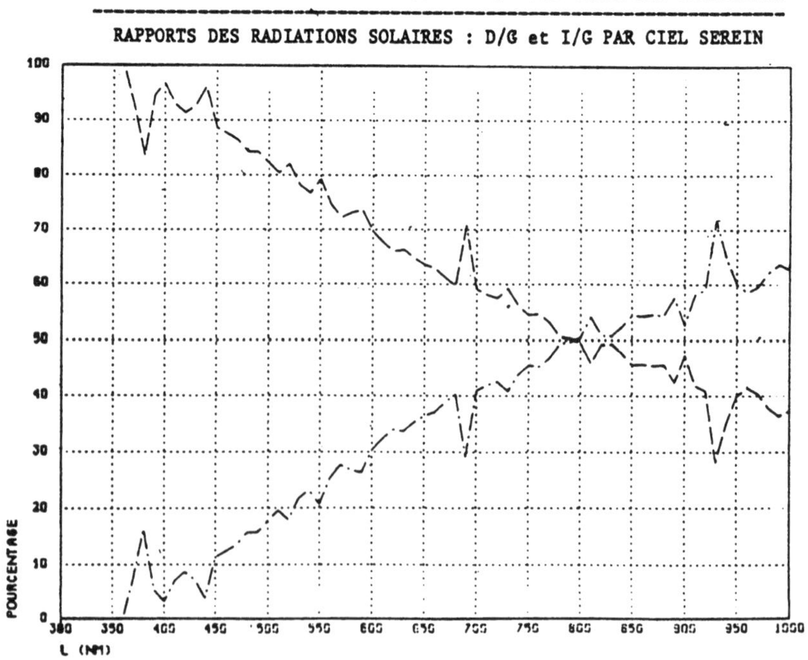

Fig. 3

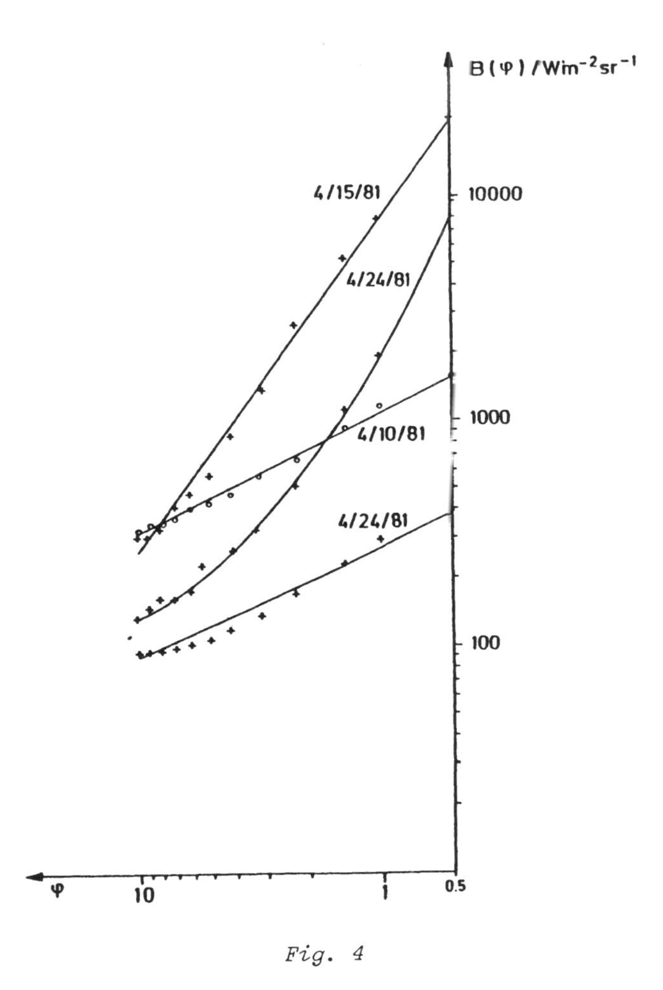

B(φ) /Wm^{-2}sr^{-1}
10000
1000
100
4/15/81
4/24/81
4/10/81
4/24/81
φ
10
1
0.5

Fig. 4

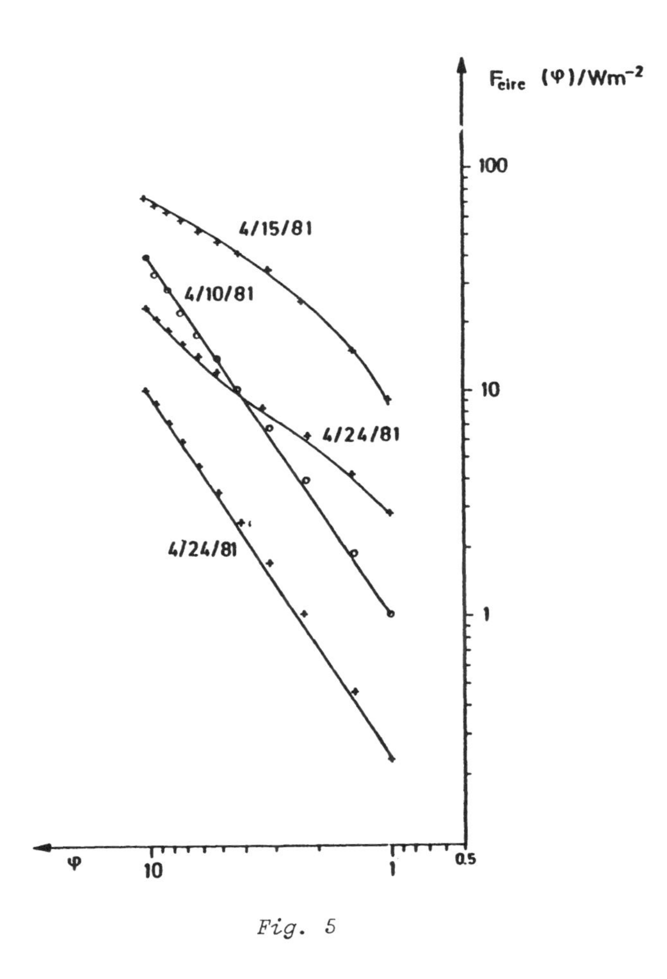

F_{circ} (φ)/Wm^{-2}
100
10
1
4/15/81
4/10/81
4/24/81
4/24/81
φ
10
1
0.5

Fig. 5

Action 4.3 - SATELLITE IMAGE PROCESSING

Action leader : Mr R. Michaël POISSON
Centre de Télédétection et d'Analyse des Milieux Naturels
Sophia-Antipolis
06560 Valbonne (France)

Participants : Pr RASCHKE - W. MÖSER (ESF 013 D)
University of Cologne
Institut für Geophysik und Meteorologie

E. REINHARD (ESF 016 80 D)
University of Stuttgart
Institut für Physikalische Electronik

J.M. MONGET (ESF 008 F)
Centre de Télédétection
et d'Analyse des Milieux Naturels
Ecole des Mines
Sophia-Antipolis

Task : Development of operational methods using satellite images for the determination of cloudiness, sunshine duration and irradiance.

SUMMARY

Action 4.3 of project F is dedicated to the study of ground radiation using satellite imagery. Within this frameworks, three groups are developing original methodological approaches either on a purely statistical basis or using realistic physical models. Work carried out over the past year is described in this paper.

1. Introduction

Action 4.3 of Project F is concerned with the determination of radiation measurements using satellite image processing. The methods investigated by the three participants have a common goal but differ as regards the types of data used and the methods themselves.

Section 2 of this report gives a succinct outline of the work of each participant and the main features of work carried out since the last meeting.

Section 3 is entitled "Joint activities in Action 4.3" and is mainly devoted to an action meeting which was held in Stuttgart in October of this year. During this meeting, requirements were reviewed both for satellite data and ground truth measurements.

2. Participants

2.1. University of Cologne

The objectives for the period under consideration were essentially improvement and detailed testing of a method based on a combination of satellite data and radiative transfer measurements. Figures I (a - b) show the result of comparisons between pyranometer measurements and satellite estimates. Figures II (a - c) are different global radiation charts obtained from satellite estimates and Figure III is a map of the DWD stations used in the experiment.

2.1.1. Preprocessing of raw satellite data

- 74 METEOSAT images (01.06.79 - 15.06.79) have been preprocessed (images have been navigated, statistical parameters heve been derived)

- The preprocessing of the next 15 days period (16.06.79 - 30.06.79) will start in December using modified routines.

2.1.2. Computation of global radiation and cloud parameters

- Amount of cloudiness N_R

  effective cloudiness N_{eff}

  global radiation M_G

  are to be computed on a 20-30 km grid (average of 6 lines by 8 columns of the METEOSAT VIS-channel).

- According to the results of the ground truth comparison, some smaller corrections will have to be applied to the model (e.g. in the case of broken cloudiness).

- The routines will be programmed more economically in early 1982.

2.1.3. Ground truth comparison

- 19 stations of the DWD (hourly data) have been used for a first comparison ; spatially interpolated satellite estimates have been compared with temporally interpolated pyranometer measurements.

- Work has been started to improve the results of ground truth comparisons ; the different influence real clouds have on the fields of direct radiation and diffuse radiation will be taken into account : the vector between the local position of the cloud in the image and the disturbance of the field of global radiation will be computed ; this will reduce the errors due to cloud geometry especially in the case of broken cloudiness.

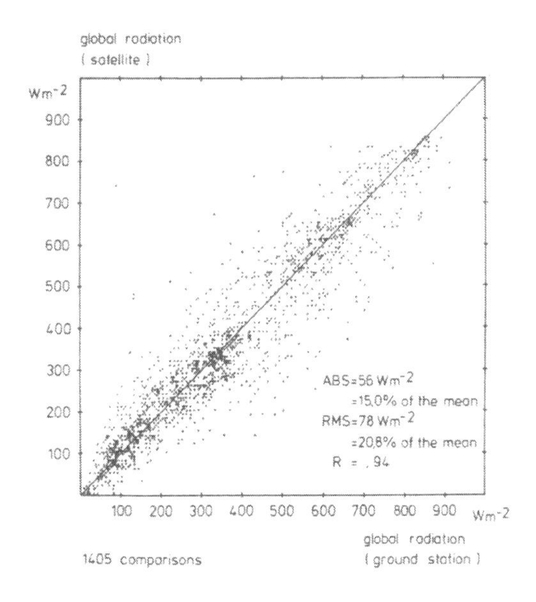

<u>Figure I.a</u>

Satellite estimates of global radiation compared to ground station measurements (up to 6 estimates per day in the period 01.06.79 - 15.06.79) have been compared with hourly means of pyranometer measurements from 19 network stations (DWD)
See also Fig. II.d

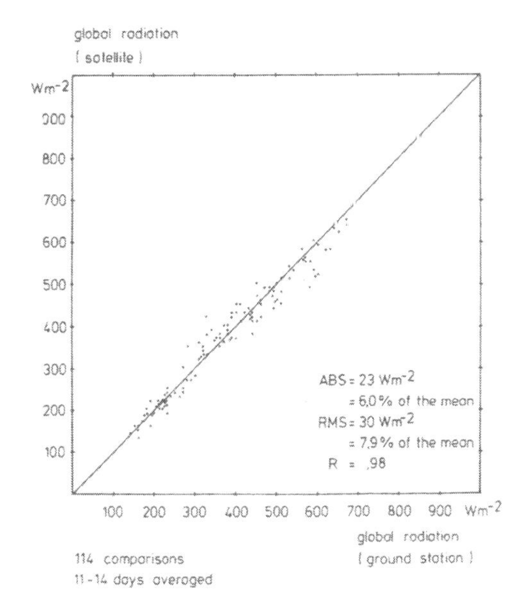

<u>Figure I.b</u>

Same as in Fig. I.a, except that 15-day averages of hourly means have been used for the comparison.

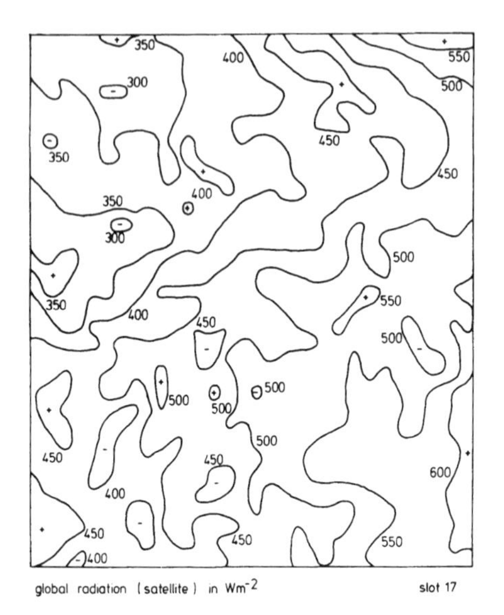

Figure II.a

Chart of 15-day averages of hourly means of global radiation :
Slot 17 (satellite image taken ~ 08.23 GMT)

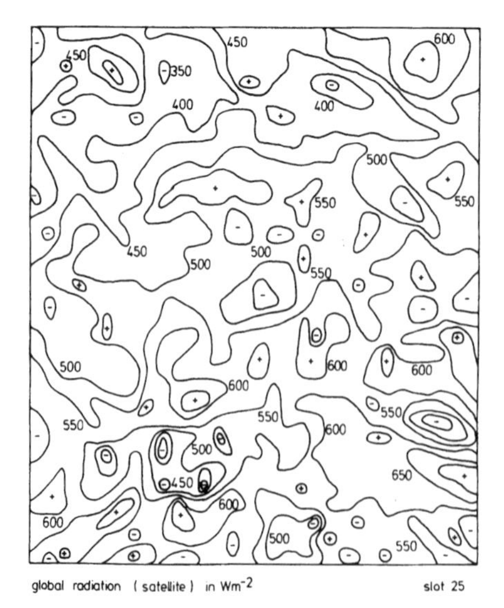

Figure II.b

Chart of 15-day averages of hourly means of global radiation :
Slot 25 (satellite image taken ~ 12.23 GMT)

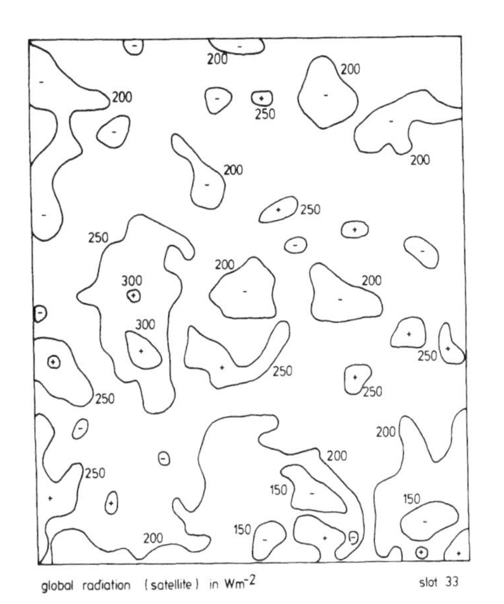

Figure II.c

Chart of 15-day averages of hourly means of global radiation :
Slot 33 (satellite image taken ~ 16.23 GMT)

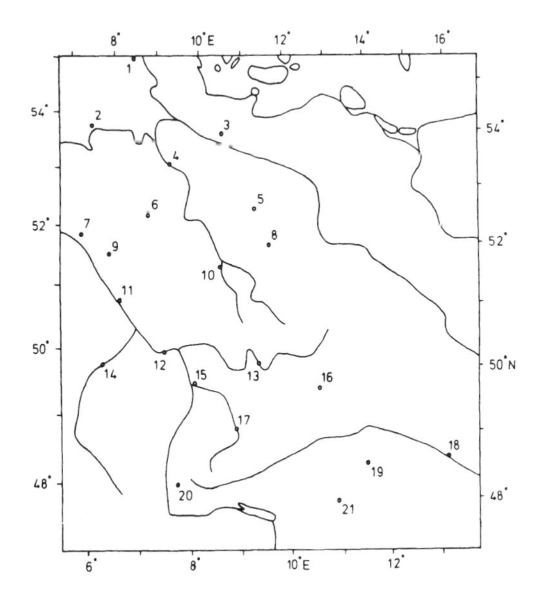

Figure III

Station chart of the DWD (31.12.79)

(Station 16 and 17 could not be used for the comparison of Fig. I)

2.2. University of Stuttgart

2.2.1. Experimental results

During the previous periods of the project, the process steps of rectification, cloud detection and cloud structure detection have been tested with a test data set of ca. 50 NOAA 5 satellite scenes. For this purpose, existing statistical classification methods have been adapted to the estimation of global radiation from satellite images.

This experiment has been carried out with ground truth data (hourly means) obtained from the Deutscher Wetterdienst Hamburg for the stations Hamburg, Bremen, Norderney, Sylt. All satellite scenes have been received in the morning at 9h ± 15 min. The feature set needed for the classification experiment was based on a 25 km area surrounding the meteorological stations mentioned above. In all 116 features evaluated from the visible and infrared image corresponding with each ground truth value have been extacted. Because of the small data set of satellite images (ca. 50 scenes in 1977 and 78) only 8 of these features could be used because of statistical reasons. The distribution of these scenes over the seasons is outlined in figure IV.

In the classification experiment, the global radiation has been divided into 4 classes :

Class 1	:	0 - 63 Joule/cm^2.h
2	:	64 - 127
3	:	128 - 191
4	:	192 - 255

The results of this experiment are outlined in the following confusion matrix of table I.

It is obvious that the classifier has the best results extreme radiation values whereas in the intermediate range the results have to be enhanced further. Within the classification process not only the first choice values are available as shown in the matrix but also information of the second and third choice which gives criteria for further interpretation of each classification. the aim until here was the test of our complete strategy of satellite image analysis.

Without changes in the strategy and program packages it is possible to introduce more information in the evaluation process. This will be done with the introduction of NOAA 6 and 7 image material containing much more information than NOAA 5 data. The improvements of the data material will be :

- Data material is available 6 times a day
- High resolution data are available in 5 different spectral channels :

0.58 - 0.68 um
0.72 - 1.10 um
3.55 - 3.93 um
10.3 - 11.3 um
11.5 - 12.5 um

- Additionally these satellites provide low spatial (ca. 40 x 40 km) but spectral very high spatial resolution data (20 channels within 3.76 and 14.95 um). These channels have special sensitivities for example to watervapour, ozone and special layers of the atmosphere. They can be calibrated with reference data from the satellite.
  A set of some of these data is shown in figure V. It is obvious that these data contain a lot of additional and from one another independent information of the meteorological situation.

To improve the results in the next period additionally to the enlarged primary satellite data a really representative data set of satellite scenes must be processed and calibrated with ground truth data.

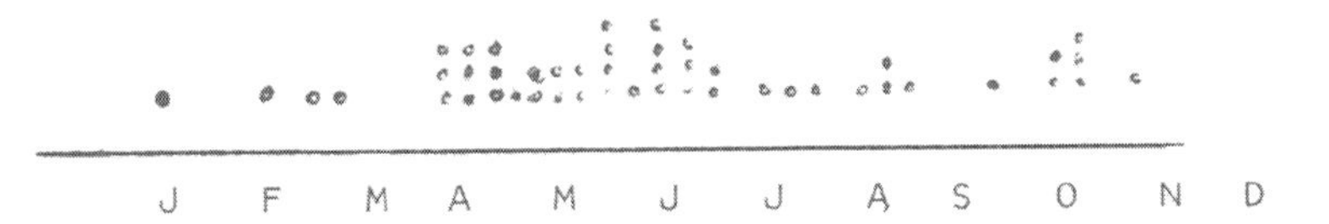

Figure IV : Temporal distribution of scenes used in the pilot experiment for the estimation of global radiation.

Ground value		Class 1	Class 2	Class 3	Class 4
	Class 1	26.14	2.27	0.57	4.55
	Class 2	12.5	9.66	0.57	1.14
	Class 3	6.25	3.41	9.09	6.82
	Class 4	0.57	1.14	2.84	12.5

Table I : Confusion matrix (in %) for classifier results versus ground observations.

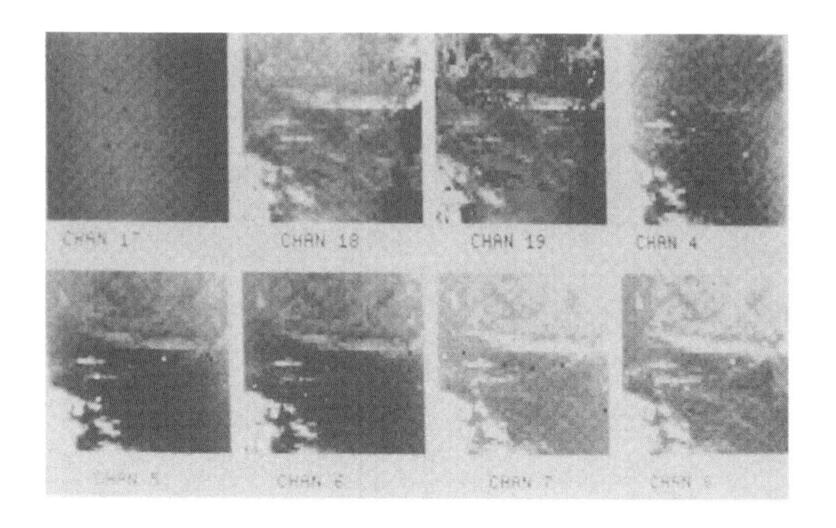

Figure V : Different spectral images received from NOAA 6.

2.3. Ecole des Mines

2.3.1. Outline

The HELIOSAT project aims to develop an operational system for global radiation estimation from satellite images. The final product will be an integrated system built round a minicomputer and low-cost (SDUS) receiving station.

The estimation will be obtained statistically from METEOSAT cloud cover measurements. Up to now, research has been aimed at the automation of accurate cloud detection in satellite images, initially on a yes/no basis, later in the form of a cloud cover index, given in percentage. This work has been carried out on PDUS (high quality) data, provided on magnetic tape.

2.3.2. Results

Since direct modelling of the measured radiance/cloudiness relationship does not give satisfactory results, time series of images are used to estimate either albedo, in the visible channel or thermal inertia for the infrared, as measured by the satellite in clear weather conditions. This estimate can be used to predict the measurements the satellite should be obtaining in cloud free conditions for each incoming image and thus to assess cloudiness.

Infrared measurements cannot be used in this scheme, because their variation is unpredictable both in time and according to the type of object under observation. The visible channel is used principally, with very good results, except in snow-covered areas (for example). The remaining ambiguities can probably be removed by using the infrared in these cases only.

Figure VI shows ground albedo for Europe computed from 10 METEOSAT images of May 1979.

Figure VII shows cloud cover indexes computed for one image in this period.

2.3.3. Equipment

The SDUS receiving station at Ecole des Mines is working since November 15th 1981. This will have two consequences on operationnal aspects :

- real time image receiving will be possible,
- the system will be independent of outside sources of data.

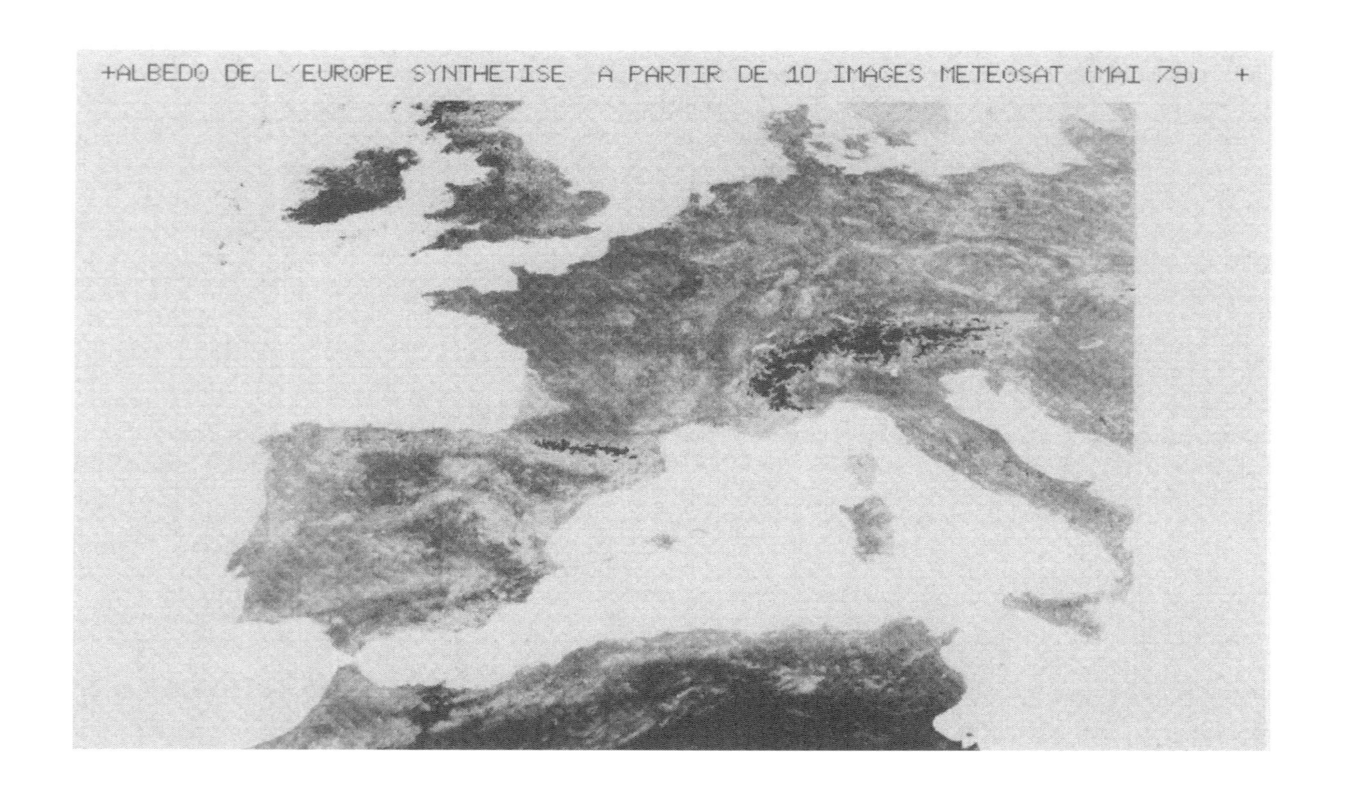

Figure VI : Albedo of Europe (May 1979)

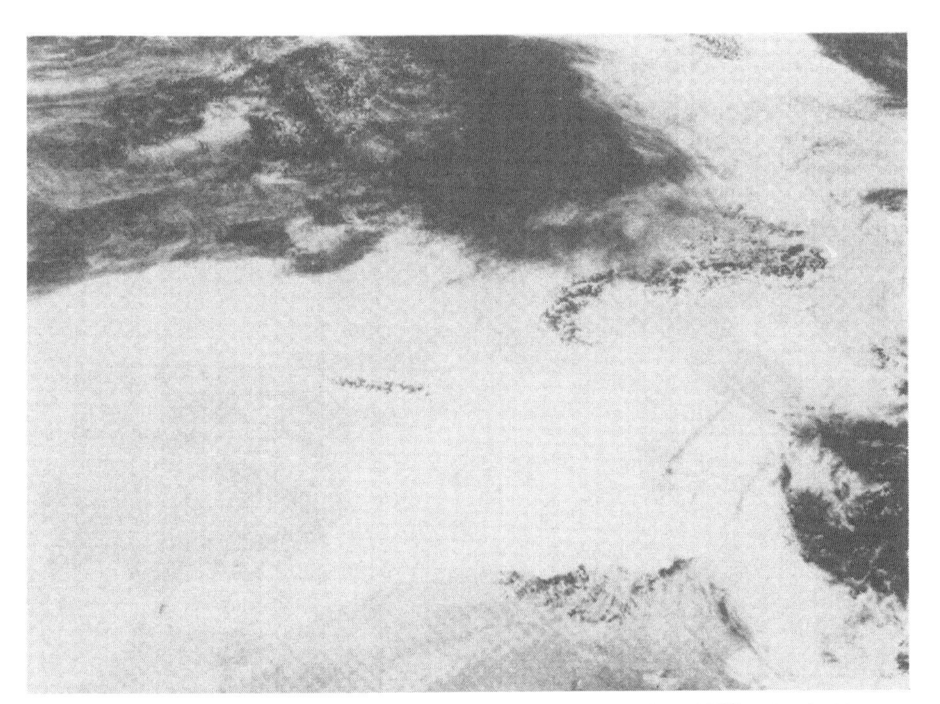

Figure VII.a : METEOSAT image of Europe, 11 May 1979, Slot 20, visible channel.

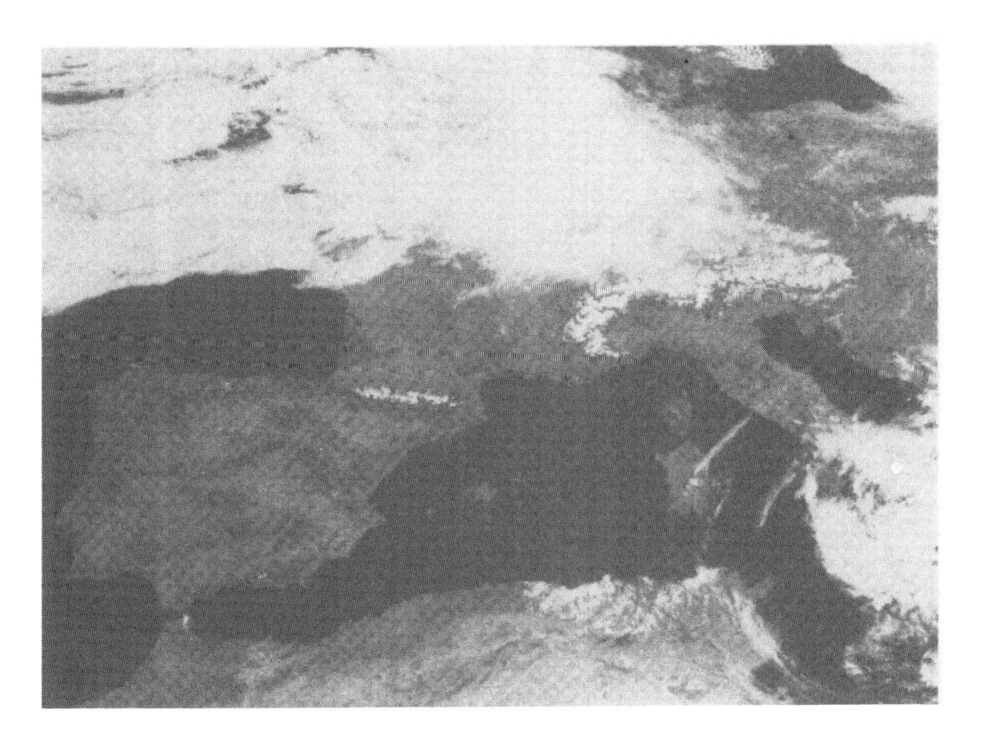

Figure VII.b : Cloudiness in % for the same image.

3. Joint activities in Action 4.3

An action meeting was held in Stuttgart on October 15/16 1981. In the first part of the meeting, the technical presentations gave rise to a thorough discussion. In the second part, two important topics were discussed : ground truth requirements and selection of a common test period.

3.1. Ground truth requirements

Up till now, participants have been acquiring their own ground truth data on an individual basis. This is inefficient in more than one respect :

- Data is usually restricted to the participant's country,
- Copying costs, when applicable, can rise quite quickly,
- Converting data to machine readable format is a time-consuming operation.

However, unification of the data collection and dissemination process requires that all participants use data sets covering identical time periods ! Although this has not always been the case in the past, the situation should improve in 1982.

Both Ecole des Mines and University of Stuttgart need data for the year 1982 : 4 x 1 month or 2 x 2 months periods in the first case, the first six months of the year in the second.

University of Cologne do not plan within their current contract to use data in 1982, but they need more data than the 19 stations currently used in June and October 1979.

From the temporal resolution point of view, it was agreed that hourly means would be sufficient. Any shorter time interval would lead to problems of availability and data handling.

Spatially, it seems that the maximum number of stations over the whole of Europe would be appreciated. This requirement must be qualified on the basis of availability and volume of data.

The possibility of obtaining data from a dense network was considered with interest as it could lead to an assessment (The spatial resolution aspects of satellite measurements).

Finally, two participants proposed to carry out any data capture (conversion from paper to magnetic tape) that may be required, as long as reasonable amounts of data were involved. Satellite results could be made available on an exchange basis, but there was some emphasis on keeping the cost of ground truth data acquisition low.

3.2. Selection of a common test period

To avoid past dispersion of effort (e.g. two participants independently selecting two consecutive 15 day periods for study in spring 79), it was agreed that future tests should be carried out on a common test period. However, differences in the type of data used as well as methodological approach may lead to some complications. More specifically, the end product of the different participants is not identical !

Nevertheless, the added efficiency of a common test period should make it worthwile, and a test will be carried out in the spring of 1982.

4. Conclusion

The main objective of action 4.3 is an operational method for estimating radiation parameters from satellite imagery, available in December 1982.

At present, one outstanding problem is the singular definition of this objective since at present there are <u>three</u> methods which differ significantly. However, each one is evolving satisfactorily from a purely experimental problem identification phase to a phase of further testing of already proven algorithms and procedures. In the near future, these tests will be carried out on identical time periods, with identical ground truth data. This should simplify any intercomparison of methodological approaches which may be ultimately required.

SPECIAL REPORT : TYPICAL DAYS

J.A. BEDEL
Direction de la Météorologie, Service Météorologique Métropolitain
2, avenue Rapp, F - 75230 PARIS CEDEX 07

1. ORIGIN AND NATURE OF THE REQUEST

The contribution of meteorologists to the design of a complex solar system is relatively poor. It must be recognized that the methods for processing meteorological data concerning the applications of solar energy are directed rather towards the size of the solar systems (storage, surface area of collectors...) than to the design itself. The method of cumulative frequency curves is a typical example : it gives the mean solar input in a solar collector without taking account of the system to which the collector is connected.

The needs expressed by researchers for studies of solar systems are quite simple. These specialists ask for files of hourly data of the main meteorological parameters for selected days, representative of the local climate.

These selected days are called : typical days.

They are used to investigate the working of the solar system, using simulation models.

The monthly and yearly frequencies of occurrence for each kind of typical day are also very useful to select a solar system adapted to local climatological conditions.

2. EXAMPLES OF 2 DIFFERENT CLASSIFICATION METHODS

The two methods consist in gathering classes of days where meteorological conditions are "similar". The typical days are then randomly selected, with one day for each class.

2.1

In the method proposed by Perrin de Brichambaut, morning and afternoon data are filed separately in 10 classes (see appendix 1).

The meteorological parameters used are the following :
- sunshine duration;
- precipitation occurrence.

For classification of the days, there are 100 possibilities (10 possibilities for the morning classification x 10 others for the afternoon classification).

This classification is simple and universal. It can be applied to any climatic condition, except for some climatic zones where it is necessary to introduce some additional criteria (e.g. haze in the Sahel..).

On the other hand, temperature data are not included in the classification. For thermal applications it will probably be necessary to subdivide the classes into sub-classes according to temperature criteria.

2.2

The "Université de Technologie de Compiègne" has developed a classification method using the objective analysis of meteorological data. This method is rather complex and requires computers; it also entails the processing of long climatological sets of daily meteorological observations. Each classification obtained by this method is representative of one station only. It is not possible to transpose a classification from one station to another.

The classes may be identified by qualitative criteria such as :
- haze and mild days;
- cold and windy sunny days.

But they cannot be defined by precise criteria or thresholds concerning meteorological parameters. Several parameters may be simultaneously considered by this method (solar radiation, cloudiness, temperature, wind, humidity..) with weights that can be chosen according to the purpose of the classification. The classifications are thus representative of a station and also of a type of solar application (photovoltaic, passive system ..).

The methods offers many advantages but it is very expensive at the present time, since it is necessary, for each specific application, to run the method on the computer with all the necessary means (data set ..).

3. DEVELOPMENTS TO BE EXPECTED IN THIS FIELD

At the present stage, definition and characterization of typical days are far from being fully settled. Further investigations are needed in this field. Among the items that need to be developed, the following may be mentioned :

3.1 The study of other classification methods.
The problem concerning classification is well known to meteorologists. Many methods have been developed for various applications. They can be divided into 4 different classes :

- classification of synoptic scale meteorological situations (e.g. Lamb's classification for the UK, Hess and Brezowsky's classification for Germany..);

- classification according to the vertical structure of the atmsphere (e.g. Pone's classification);

- classification according to some criteria or thresholds for one or several meteorological parameters (e.g. Perrin de Brichambaut's classification already mentioned);

- objective classification (e.g. the above classification developed by the "Université de Technologie de Compiègne", mentioned in 2.2 above).

For this item, it is possible to consider :

- a bibliographical study of the classification methods used in meteorology and a critical study of their usefulness for solar energy applications;
- studies of the adaptation of some classification methods for application to solar energy.

3.2 Presentation of meteorological data for the typical days.
It is possible to choose the real observations for a particular day or to set mathematical formulas giving the values of meteorological parameters according to the time (for instance : the parameterization proposed by Dogniaux. See Dogniaux's report presented at the coordination meeting of contractors, Brussels, April 1 - 2, 1981).

3.3 Daily sequences of data, representative of a local climate.

3.4 Inquiry about other methods of processing meteorolotical data, adapted to the needs of designers of solar energy systems.

4. PROJECT OF ACTION FOR GROUP F

4.1 Appointment of an action leader responsible for this project.

4.2 Organization in 1982 of a meeting with a view to :

- reviewing the running studies and projects;
- proposing priority studies in that field.

APPENDIX I - CLASSIFICATION OF DAYS FOR SOLAR ENERGY APPLICATIONS (after Perrin de Brichambaut)

TWO-DIGIT CLASSIFICATION (100 classes)

The first digit indicates the class for the morning
The second digit indicates the class for the afternoon
Each half-day (morning or afternoon) is classified as follows (in 10 classes)

CLASS	According to relative sunshine duration	According to average cloud cover (in oktas)
0	$S \geq 0.8$	$N < 1.5$
1	$0.8 > S > 0.2$	$1.4 < N < 6.5$
2	$0.8 > S > 0.2$ + rain	$1.4 < N < 6.5$ + rain
3	$0.05 \leq S \leq 0.2$	$6.5 \leq N \leq 7.7$
4	$0.05 \leq S \leq 0.2$ + rain	$6.5 \leq N \leq 7.7$ + rain
5	$S < 0.05$	$N > 7.7$
6	$S < 0.05$ + rain	$N > 7.7$ + rain
7	$S < 0.05$ + continuous rain	$N > 7.7$ + continuous rain
8	visibility less than 5 km during the entire half-day	
9	visibility less than 1 km during the entire half-day	

S_M : relative sunshine duration for morning

S_A : relative sunshine duration for afternoon

N_M : Average cloud cover for morning

N_A : Average cloud cover for afternoon

METEOROLOGICAL TERMINOLOGY

Air mass

The path through the Earth's atmosphere traversed by the direct solar beam, expressed as a multiple of the vertical path to the same point, calculated for sea level.

Solar elevation

The angle between the direct solar beam axis and the horizontal (degrees). Air mass $\approx$ cosecant of the solar elevation for elevations $\geq 10°$ and air mass ≤ 5.6.

Angle of incidence

The angle between the direct solar beam axis and the normal of the receiving plane.

Albedo

The reflectance of a surface for global solar radiation. It is the integral of the global irradiance, weighted by the spectral reflectance of the surface, normalized to the global irradiance.

Spectral reflectance

Ratio of reflected spectral irradiance normalized to the incident spectral irradiance of the same wavelength.

Power quantities

The radiant power is the power of electromagnetic radiation impinging onto a plane.

The radiant flux density is the radiant power/unit area.

Direct solar irradiance

The radiant flux density of the solar radiation received from a certain solid angle around the sun by a plane at an angle of incidence α.

The direct normal solar irradiance is received at an angle of incidence $\alpha = 0°$. It is measured by *pyrheliometers*.

Diffuse solar irradiance

The radiant flux density of scattered solar radiation (W m^{-2}), received by a plane within a solid angle of 2π.

Global solar irradiance

The solar radiant flux density (W m^{-2}), received by a plane within a solid angle of 2π; if not otherwise stated, the plane is horizontal. Global irradiance = direct irradiance + diffuse irradiance received on the same plane.

Global and diffuse solar irradiance are measured by *pyranometers*.

Note : For some years, global irradiance has been expressed by reference to a plane oriented in any direction, not only to a horizontal plane.

Spectral irradiance

The irradiance (global, direct or diffuse) per unit bandwidth at a particular wavelength (W m^{-2} μm^{-1}).

Energy quantities

The time intervals of the power quantities related to solar radiation are called irradiation (sometimes also "sums" or "insolation" e.g. "hourly sums", "global insolation", but meteorologists prefer irradiation.

Pyranometer

A radiometer used to measure global irradiance (or, with a shading device, diffuse irradiance) on a horizontal or inclined plane with a view angle of 2π steradians. If inclined, radiation reflected from the foreground is also detected; the amount of reflected radiation depends very much on the reflection properties ("albedo") of the foreground.

Pyrheliometer (also "Normal Incidence Pyrheliometer (NIP)" or in older publications "Actinometer")

A radiometer in which an aperture limits the view angle of a plane receiver. A typical aperture angle is 5.6°. That pyrheliometer is tracked in order to follow the sun, so that the direct beam axis falls perpendicularly onto the receiving plane.

LIST OF PARTICIPANTS

BEDEL, J.-A.
Direction de la Météorologie
Service météorologique métropolitain
2, avenue Rapp
F - 75340 PARIS CEDEX 07
Tel.
Telex

BUIS, H.
Commission of the European Communities
D. G. Science, Research and Development
200, rue de la Loi
B - 1049 BRUSSELS
Tel. 736 60 00 x 8444
Telex 21877 comeu b

DOGNIAUX, R.
Institut Royal météorologique de Belgique
3, avenue Circulaire
B - 1180 BRUXELLES
Tel. 02/374 67 88
Telex 21315 meteor b

DURBIN, W.G.
Observational Requirements and Practices Branch
Meteorological Office - HQ Annexe
Eastern Rd
UK - BRACKNELL, Berkshire RG12 2SZ
Tel. (0344) 20 242 x 2538
Telex

GANDINO, C.
Commission of the European Communities
Joint Research Centre
I - 21020 ISPRA (Varese)
Tel. (0332) 78 02 71
Telex euratom - italy

GRUETER, J.W.
Kernforschungsanlage Jülich
Programmgruppe f. Systemanal. u. techn.Entwicklung
P.O.B. 1913
D - 517 JUELICH
Tel. (2461) 61 33 97
Telex

KARALIS, J.
Department of Meteorology
University of Athens
33, Ippokratous street
GR - 144 ATHENS
Tel. (01) 36 10 006
Telex

KASTEN, F.
Deutscher Wetterdienst
Meteorologisches Observatorium Hamburg
Frahmredder 95
D - 2000 HAMBURG 65
Tel. (040) 601 79 24
Telex 2162912 dwsa d

KOCH, R.
KFA Jülich
Institut für Kernphysik
D - 5170 JUELICH
Tel. (02461) 61 41 20
Telex

LUND, H.
Technical University of Denmark
Thermal Insulation Laboratory
DK - 2800 LYNGBY
Tel. (02) 88 35 11
Telex 37529 dthdia dk

McWILLIAMS, S.
Valentia Observatory
Cahirciveen
IRL - CO KERRY
Tel. Cahirciveen 27
Telex 26912 mtva ei

MURPHY, E.
Irish Meteorological Service
Valentia Observatory
Cahirciveen
IRL - CO. KERRY
Tel. Cahirciveen 27
Telex 26912 mtva ei

NICOLAY, D.
Commission of the European Communities
D.G. Information Market and Innovation
P.O.B. 1907
L - 2920 LUXEMBOURG
Tel. 43011 x 2946
Telex 3423/3446 comeur lu - 2752 eurdoc lu

PAGE, J.
University of Sheffield
Department of Building Science
Western Bank
UK - SHEFFIELD SIO 2TN
Tel. 0742 78555
Telex

PALZ, W.
Commission of the European Communities
D.G. Science, Research and Development
200, rue de la Loi
B - 1049 BRUSSELS
Tel. 735 00 40
Telex 21877 comeu b

PLAZY, J.-L.

Commissariat à l'Energie solaire
Sophia Antipolis
F - 06560 VALBONNE
Tel. (93) 74 79 79
Telex comessa 461357 f

POISSON, M.

Centre de Télédétection et Analyse des milieux naturels
Ecole des Mines
Sophia-Antipolis
F - 06560 VALBONNE
Tel. (93) 33 05 58
Telex

SLOB, W.

Royal Netherlands Meteorological Inst.
Wilhelminalaan 10
NL - 3730 AE DE BILT
Tel. (030) 76 69 11
Telex

STEEMERS, T.

Commission of the European Communities
D.G. Science, Research and Development
200, rue de la Loi
B - 1049 BRUSSELS
Tel. (02) 735 00 40 x 6878
Telex 21877 comeu b

VIVONA, F.

Consiglio Nazionale delle Ricerche
Istituto Fisica Atmosfera (IFA)
Piazzale Luigi Sturzo 31
I - 00144 ROMA
Tel. (06) 59 17 625 / 85 43 83
Telex 612322 cnr pfe i / 614344 atmos i